Home Theaters
And Electronic Houses

Home Theaters

And Electronic Houses

CEDIA

and

Tina Skinner

Library of Congress Cataloging-in-Publication Data

Home theaters and electronic houses / by CEDIA and Tina Skinner.-- Large print ed.
p. cm.
ISBN 0-7643-1957-4 (Hardcover : large print)
1. Home entertainment systems. 2. Large type books. I. Skinner, Tina. II. Custom Electronic Design and Installation Association.
TK7881.3 .H68 2004
621.338--dc22
2003018885

Designed by Bonnie M. Hensley
Cover design by Bruce Waters
Type set in Beauty School Dropout/Zurich LtCn BT

ISBN: 0-7643-1957-4
Printed in China

Published by Schiffer Publishing Ltd.
4880 Lower Valley Road
Atglen, PA 19310
Phone: (610) 593-1777; Fax: (610) 593-2002; E-mail: Info@schifferbooks.com
Please visit our web site catalog at **www.schifferbooks.com**
We are always looking for people to write books on new and related subjects. If you have an idea for a book, please contact us at the above address.

This book may be purchased from the publisher. Include $3.95 for shipping. Please try your bookstore first. You may write for a free catalog.

In Europe, Schiffer books are distributed by Bushwood Books
6 Marksbury Avenue
Kew Gardens
Surrey TW9 4JF England
Phone: 44 (0) 20 8392 8585; Fax: 44 (0) 20 8392 9876; E-mail: Bushwd@aol.com
Free postage in the UK. Europe: air mail at cost.

CONTENTS

FOREWORD

CEDIA®

CUSTOM ELECTRONIC DESIGN & INSTALLATION ASSOCIATION

A home theater is unique. Unlike other rooms in the house, a home theater allows us to use the absolute best of our products. It tests the limits of our imaginations. It delivers a level of performance and an experience that simply cannot be recreated in any other place. It is our escape, our entertainment and passion. Home theater truly brings magic into the comfort, safety, and convenience of our homes.

The Custom Electronic Design and Installation Association (CEDIA) was founded by dealers installing home theater systems designed with the highest performance audio and video products available. CEDIA has evolved to become a global association of companies that specialize not only in high-end home theaters, but also in the design and installation of all high quality electronic residential systems.

We are your advocates. CEDIA is to the homeowner as the American Institute of Architects is to the architect or as the American Medical Association is to the medical community. Our member companies are established and insured businesses with bona fide qualifications and experience in this specialized field.

CEDIA's emphasis has been on the education and certification of our members. We don't simply teach the nuts and bolts of an electronic installation; we teach our member companies how to handle every aspect of their business. Our members go to "Electronic Lifestyles™ School" so you don't have to. CEDIA's designs and installations meet the requirements, needs, and wishes of the entire family.

CEDIA and its members are dedicated to helping you create beautiful, high performance, fun rooms just like the ones you are about to see. We hope you will enjoy the fascinating rooms in this book as much as we enjoyed creating them.

If you would like to find out how you can experience a true Electronic Lifestyle™, please contact CEDIA at 800-669-5329 or use our online Finder Service, www.cedia.org\finderservice, to find a custom installer near you.

Jeff Hoover
President
Custom Electronic Design and Installation Association

Acknowledgements

Editor: Tina Skinner

Project Coordinator: Lindy McCord

CEDIA Liaison: Brittany Nims

Writers: Tina Skinner
Lindy McCord
Jeff Hoover
Mitchell Klein
Ed Condiracci

The CEDIA members whose work is shown are the true talent and inspiration behind this beautiful book. Special thanks also goes to Billilynne Keller, whose vision and decision made this book finally spring to life, and to Carol Mattson, working behind the scenes to push everyone along.

INTRODUCTION

Welcome to fantasy world, except this is real life, real homes.

The amazing homes featured on these pages are packed with Wizard of Oz effects, with curtained and cabinet-encased stacks of computerized electronic gizmos that bring Hollywood fantasies to life.

At a one-touch command on a wireless panel, you can dim the lights, draw the drapes, and lower the screen and projector from the ceiling. A well-designed room can exceed the performance of even the best public movie theaters.

Behind the scenes, a lot of know-how and technical expertise went to work creating leisure and comfort through state-of-the-art audio, video, lighting, communications, security, and control equipment. Real-life Wiz kids – professionally trained and certified installers – made it all possible, in conjunction with interior designers and architects who have tapped the code to unlock illusions and grandeur.

A home theater may cost as little as $20,000 in gear – a small projector, surround sound, etc. However, they more commonly start at $45,000-70,000, and prices can easily soar over $1 million.

Nonetheless, with a trend toward staying home, many people are investing in home entertainment instead of travel abroad. Home theaters are the hottest trend in housing today. They are the perfect solution to the unfinished basement, where masses of residential square-footage have long gone to waste. Relatively soundproof by its very underground nature, a basement is the logical location for a two-hour (feature film) escape. Today's architects work with installers to retrofit an existing basement, or to design new basements with theaters in mind. Sloped floors allow for staggered seating and raised ceilings create clearance space for projectors and speakers. For homes where basements aren't an option, experts can soundproof other rooms, create retractable covers to conceal windows, and make any other necessary adjustments to ensure an excellent media environment, be it dedicated or multi-purpose. For homeowners who will settle for nothing but the best, computerized acoustic diagnostics, box-within-a-box construction, subflooring, insulation, and sound panels are among the many components technicians utilize to ensure soundproofing for other parts of the home and perfect sound within the theater.

For those who love movies, having the big screen at home eliminates the hassle of crowds, parking, lines, and inconvenient show times. The movie is ready when you are, and pauses when you're not. Plus, you aren't fighting someone for the armrest. An entire furnishing industry has been created to meet the demand for plush, ergonomic theater-style seating. Folks with home theaters have the capability to invite the entire family to view a movie or to invite clients to attend a private presentation.

The following essays come from professionals within the industry, and will lend insight to homeowners looking to create their own magical theater experience. A glossary of terms will help define any questionable language, and the gorgeous photo gallery will delight and inspire!

How to Hire a Custom Installer

By Mitchell Klein

Finding a custom installer to design your home entertainment systems may seem like a daunting endeavor. You're dealing with consumer electronics, after all, and electronics can be intimidating. Yet hiring a custom installer is really no different than hiring an architect, interior designer, or other professional. The most important factors to consider are experience and connecting with someone you can trust – someone who isn't going to "drop the ball" halfway through the project.

There's no need to feel intimidated, either, because an experienced custom installer recognizes the importance of establishing a good rapport with clients from the start. If the installer has an "attitude," don't waste your time, because the attitude will only get worse. Your future custom installer should make you feel relaxed and appreciated, not patronized. Any lack of knowledge about electronics on your part should be met with an enthusiastic and sincere desire to educate you. The installer should be willing to take the time to help you understand the options available to meet your requirements and your budget.

Before you begin your search, ask yourself these fundamental questions: Are you looking for the best firm with complete, beginning-to-end responsibility for your project? Is this firm expected to design, engineer, provide comprehensive documentation, and integrate with all the other parties involved (i.e., architect, interior designer, builder, cabinetmaker)?

Perhaps you're looking for the lowest possible price or most economical solution to your needs. Make sure you understand the difference between "simple" systems (not necessarily easy to use or upgrade in the future) and "simple to use" systems. A "simple" system will utilize some basic equipment and will require the user to understand how to operate all the individual components. A "simple to use" system incorporates complex products designed to integrate with reliable control systems to enable the user simple, predictable operation. A "simple" system will cost far less than a "simple to use" system; but may actually rarely be used at all due to frustration with operation. With custom installed home media systems, the old adage "you get what you pay for" applies in spades.

The following guidelines will help prepare you in your search for the technical "guru" you'll be relying on for years to come. As with any professional relationship, make sure you look for the right fit between you and the installer. This isn't like having appliances delivered and hooked up by someone you'll never see again. If you've chosen well, you'll be in contact with your custom installer for many years to come ... and loving it!

Check references of comparable projects

It amazes me how seldom clients ask for references, much less contact them. A minimum of three should be obtained (and make sure they're not friends or relatives of the installer). Request that these references represent systems of comparable scope to yours. It's of limited value to discuss an elaborate lighting control installation with a reference who merely bought a TV from this firm. Ask references the following questions: Are they happy with the system? Are they satisfied with the work performed? Can they actually operate the system with ease? Was the installer responsive to their needs and requests, or did he or she constantly cajole them into altering their expectations? With hindsight, how would they have done things differently, if at all?

When obtaining references, I recommend acquiring at least three from the "trade," ideally the architect, interior designer, and cabinetmaker, if applicable. Whether or not you're working with one is unimportant. An installer's ability to work closely with a design professional is a good barometer of his or her professionalism.

Did the installer keep the project on schedule to the satisfaction of everyone involved? Would they recommend the installer to their own clients? Was the installer instrumental in preventing or overcoming problems through good communication? Did the installer contribute innovative ideas and solutions? Most important of all, were their end result expectations met or exceeded?

Inspect the facility

If a picture is worth a thousand words, a visit to the installer's office is worth thousands of dollars. A visual inspection of the installer's operation will speak volumes about their work. Granted, not every installer will have an extensive display or demonstration facilities. All you want to ensure is that this firm is neat, organized, and committed to its business. The staff's experience and service, attention to detail, and business management is indicative of their ability to service your needs today, and tomorrow.

Trust your instincts. Call the office. Do you get a "live" person, or an answering machine? If it's a machine, do they return your call promptly? Can you reach your contact when you (or other professionals such as cabinetmakers and interior designers) need to? Does the organization appear to be organized? Do you get a good feeling when you tour the facility?

You may well hit it off with the salesperson and get all the necessary "warm fuzzies." This is certainly important. Of greater importance, however, is meeting those who will be responsible for the implementation of your project. Insist on meeting the project manager who will be assigned to your project. This is really the "can do" individual you will be working closest with. Meet other staff members as well to get an overall sense as to their commitment to servicing you for the long term.

Whose products do they sell?

I must caution you about choosing a custom installer solely on the basis of the brands they carry. Examine the installer's overall product mix and ask why certain suppliers were chosen. The answer can be very revealing about the company. Be wary if the installer says: "I'm not a dealer for their products, but I can get them." Rest assured that most manufacturers choose their dealers carefully. Make sure the installer is a factory authorized dealer for those products that are vital to your system. If a product needs repairing under the warranty, you don't want to have to send it to some warehouse in another state. You'll want your custom installer to take care of it and offer a temporary replacement if necessary. Another important point regarding authorized dealers concerns prices. Installers cannot carry every brand on the market. They select specific manufacturers' goods based on their own preferences, the availability of those products within their region, the reputation of the manufacturer for reliable servicing, and competitive pricing.

Determine their credentials and industry knowledge

It amazes me just how many people have become overnight "experts" in so many disciplines so quickly. I run into associates all the time who have expanded into home theater or lighting control systems, after years of installing car stereos, burglar alarms, or maintaining corporate computer networks. When it comes to custom home entertainment installations, there is nothing more dangerous – or costly – than a "little knowledge." Granted, in this business of rapidly advancing technologies, we're learning all the time. Just make sure your installer isn't learning at your expense.

Interview your prospect. Ask what his or her educational background is. Why did they become interested in custom installation work? Where were they previously employed? Was it an electronics store, another installer, an alarm company, an aeronautics firm? Ask about their views on audio and video and future technological advances for the home. Ask about their commitment to really learn this business.

The electronics systems industry is changing all the time; what you know today is only a foundation for what you'll need to know tomorrow. Continuing education is essential for any custom installer who plans on delivering satisfied customers for a very long time. Fortunately, there are now some credentials to judge the commitment your potential provider has made to the their employees and customer.

The Custom Electronic Design & Installation Association (CEDIA) offers courses throughout the year on a wide variety of topics – from wiring and acoustics to home automation, future technologies and business practices. Ask your prospective installer if he or she is a CEDIA member and if they've attended any of these seminars. This is not to imply that only CEDIA members are competent installers, but it's a good starting point.

CEDIA now offers certification of individuals for a variety of competencies. Certification only applies to individuals. A CEDIA-member company is NOT certified – their people must be! This is a major commitment the CEDIA-member firm has to make to its employees and customers – both of a financial nature and a time commitment. Field technicians may now acquire an Installer Level I (base level) and Installer Level II (higher level) certification. This qualification ensures the individual working in your home has committed to a process of ongoing education and achieved measurable skills; and that the company he/she works for has committed their resources to build competencies in the field. How many certified installers does the company you're interviewing have on staff?

CEDIA also has begun a certification program for system designers, which ensures a measurable competency in the design and configuration of the very systems you are considering installing in your home. Additional certification classifications are under development. Make sure to ask about the commitment they are making to the education of their people!

Once again, trust your instincts. If a "deal" seems too good to be true, it is. Remember, you're buying the entire company when you purchase an installed system. Look at the company as a whole and make your decision based on your impressions of everyone who works there.

When asking questions, don't try to intimidate the installer; but don't be afraid to ask tough questions. If their answers are vague, be skeptical. In addition to experience, you're primarily interested in their enthusiasm. They should be passionate about what they do.

You will have a long-lasting relationship with your custom installer. Make sure the company you choose uses good business practices and demonstrates their capabilities to be there for you when you need them for the long haul.

Checklist

• Verify that the installer is fully insured and licensed (in states where it's required). Do they have CEDIA certified employees on staff?

• Get quotes in writing. Make sure it is an all-inclusive, closed end, firm bid. Be clear on all and any additional charges.

• Make sure you fully understand what each comparative system is designed to do, how it will function and how it will perform. No two systems are ever exactly alike.

• Never tell one installer what another installer has bid, because if they lower their bid, you can bet that you'll pay for it somewhere – either in equipment, time or service. And be cautious about the lowest quote. You're buying a complete products/service/performance package. The lowest price is not necessarily the best option.

• Make sure you understand exactly which responsibilities the installer is accountable for. Does the quote include system pre-wiring? Documentation? Elevations? Wiring diagrams? Project meetings and management?

• Procure a written inventory of all equipment: exact brand, make and model numbers of every product you are buying. Accept no unauthorized substitutions.

Mitchell Klein is President/CEO of Media Systems, the award-winning custom installation firm headquartered in Boston with an office in Palm Beach (www.mediasystems.com). He is a two-term past President for CEDIA, and continues to serve on the Association Board of Directors.

(Reprinted from *Audio/Video Interiors*, ©1995 Avcom Publishing, Ltd.)

Behind the Scenes

What a "Best Dressed" System Means

By Ed Condiracci

All systems installed by Electronics Design Group (EDG) have an eye toward the future. They are designed to be compatible with current and future technologies, as well as installed in a manner that is durable and can grow with the home. While many electronic components have an expected lifespan of less than twenty years, most of the homes we work in will stand for one hundred years or more. Our systems, like all electro-mechanical systems in the home, must be easily serviced, upgraded, or replaced.

A "best dressed" system is one that is installed very neatly, with all cabling tied down and clearly labeled. We also make an effort to minimize the amount of field wiring that enters the equipment racks by terminating many of these cables in distribution boxes on the walls of the utility room. This keeps the racks very clean and makes them easier to service or upgrade in the future. It also allows us to better meet the timelines of an individual project. For example, on a "normal" project timeline, a lot of the field terminations are done months before the equipment racks arrive on site, but on a "tight" timeline we can have a crew on site terminating field cables, while another crew is at our shop assembling and testing the equipment rack(s). This flexibility also helps us to control the quality of the installations by all but eliminating the temptation for anyone on our staff to rush through an installation. So you won't see thousands of multi-colored wires entering the rear of our equipment racks, because those wires would have to be terminated by one installer over the course of days or weeks, not to mention the amount of clutter and difficulty they cause to later service and upgrade. Instead, we create very clean, durable installations that are done by a team of top industry professionals.

Ed Condiracci is the vice-president of production for Electronics Design Group in Piscataway, N.J., www.edgonline.com.

A network of color coded veins pumps life into the electronic heart of this home. If the installation is not done cleanly and professionally, following the road map of the digital system could be nightmarish.

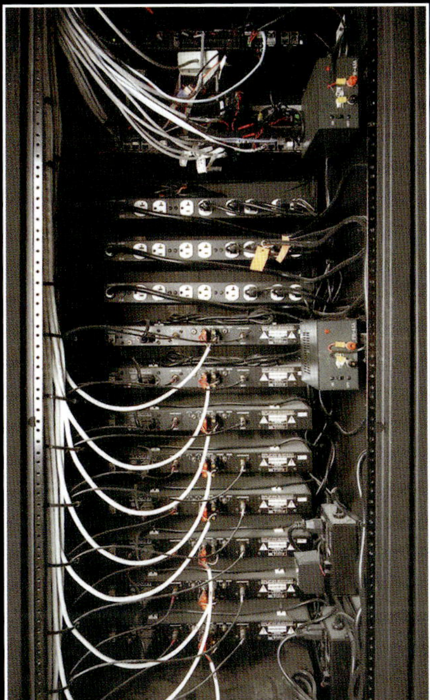

Courtesy of Digitech Home Systems, Bellevue, Washington

Courtesy of Digitech Home Systems, Bellevue, Washington

Courtesy of Digitech Home Systems, Bellevue, Washington

One would never know from the inside of a theater the extent of wiring and electronic routing that is necessary to run an audio/visual system. With the twists and curves of the wires smartly concealed and minimized, an otherwise unsightly tangle of cable becomes a tidy rainbow of digital power.

The Electronic Home

Enter a world where a mere index finger wields complete control. From the front door to the back, inside lights to lawn illumination, everything is wired into central control units and activated via remote controls or stationary monitors located throughout the home. A central unit may contain hundreds of DVD or CD units, all of which are accessible from any number of rooms simultaneously. Temperature control and security monitoring are additional features that can be included. Moreover, these complex, interconnected systems are so simple to operate even a child can master them. Anything is possible today in home electronics, as the following award-winning homes demonstrate.

Flat of the Future

A young man, single and a top-manager in banking, wanted his flat wired for state-of-the-art on and off time. He wanted a high-quality home theater, acoustic equipment and video displays in all rooms, electronic control of lighting and curtains, video monitoring, and a big-monitor personal computer wired for the Internet.

The apartment made locating a home theater zone and the electronic equipment quite challenging, and the acoustics of the premises were bad. The solution was to utilize two supporting columns to build a triangular room where all the electronic equipment, the projector, lighting, and curtain control systems were allocated. One of the walls of this room was used for allocation of the theater equipment (the screen and front and center speakers). The second wall was used for allocation of lighting and curtain control equipment and for stabilizers. And the third wall was used for allocation of the sources, acoustical equipment, and for transmission of video picture from different sources.

In all rooms and zones, touchpanels, remote controls, or push-button wall panels were installed, so the resident can control lighting, curtains, blinds, acoustical equipment, video, and video monitoring from any place in his apartment. In-wall speakers and displays were installed in every room so the resident is able to listen to music and see images from any source, including the bathroom and kitchen. The TV set that was installed in the bathroom had to be protected from high humidity and water splashes. To do that, special ventilation was built into the wall niches to divert heat.

In the anteroom, there is a wall panel to control all equipment in the apartment upon entering or leaving.

Installer: En-Trade, Russia

Showplace

This homeowner is a high-end, custom residential builder who uses his personal home as a model home for prospective clients. Therefore, the house had to be accessible to guests at a moment's notice, and flexibility and convenience were of paramount importance. A college-aged daughter and an elderly aunt live in the home, too, so the system had to be easy to operate for three generations.

The room that was designated for the theater also doubles as a hurricane safe-room. This room has solid concrete poured walls and is octagonal in shape. All of the systems were designed for use with a back-up generator as an alternate power source.

The management of all of the home systems was integrated with an AMX system. These sub-systems include lighting control, multi-room audio, a family room surround system, a residential theater system, motorized hurricane shutters, motorized bug screens, and control of the many water features. An in-wall touchpanel was placed in the kitchen and the foyer, and a tabletop panel was located on a bedside nightstand in the master suite. The theater and the family room audio/video system are controlled via hand-held touchpanels.

The lighting control system, with its 5 panels and 53 lighting control keypads and engraved buttons, simplifies the control of over 400 feet of cove lighting and 250 recessed light fixtures. The time clock feature in the system turns on specific lights every evening to welcome home the family.

The whole house audio system is controlled through the lighting keypads via the backbone of the home automation system. This design promotes the aesthetics of consistent keypad style with the added flexibility of expanding the audio control to any of the lighting control keypads.

The one-button control of the motorized bug screens and the motorized hurricane shutters allows the occupant shelter from nature's elements and affords extended use of the rear lanai outdoor living space to enjoy the plasma mounted above the outdoor fireplace.

When it's time to relax, the theater's touchpanel adds to the enjoyment of the automated audio/video experience through a simple button press.

Through any of the conveniently placed touchpanels, the home automation system allows the homeowner, on a moment's notice, to transform the entire residence for "wowing" guests.

Installer: AVL Pro, Inc., Naples, Florida

House of Toys

This homeowner came from a computer background and had an existing interest in home automation and a passion for electronic toys. His strong interest in the world of home integration, and a solid integrated technology background, made the project a little easier for the professional installers.

The design of this system was not difficult, even though it was extensive. The house was broken up into two systems: a digital theater system including projector, screen, and digital sound system, and the rest of the house.

Full control of all systems, i.e. lighting, audio and video, security, heating, ventilation, air conditioning, etc., is available in all eighteen rooms. The audio distribution is provided by an MP3 server, and at any one time up to fifty percent of the rooms can have different sources playing without interfering with anyone else's enjoyment. Video distribution is provided by a 700-disc DVD player, playing up to four movies at a time into different zones. Digital TV and Pay TV sources are also distributed throughout these eighteen zones.

Control is via LCD panels in each room along with a couple of roaming RF two-way panels for outdoor use. The LCD panels display the basic information for day-to-day living, but also allow the homeowner to see what movie his kids might be watching, or view the whole house plan on the screen to check room temperatures, etc.

The lighting was integrated into the main control system. Each room has the option of a plasma, TV, or projector with flush in-ceiling speakers providing the audio.

In each of the two children's rooms, there is an input panel allowing easy connection of computer games and other outside sources into the main system without ugly cables hanging down from monitors. The indoor swimming pool was wired for a plasma screen and LCD control panel, plus security and cameras, and there are speakers in the children's playground, barbeque area, and the tennis court.

Installer: Len Wallis Audio, Australia

oz top free alternative singles
2 fall down
jebediah

Media Saturation

This system was designed so casual users, such as friends, family, and business associates, could easily access all of the home's systems. For this busy family of five, it was essential to have an easy-to-use home automation system for day-to-day activities, as well as for frequent entertaining in both the main living areas and in the expansive wine cellar. As with all systems within this highly detailed environment, the door locks were designed to fit within the home, constructed primarily of concrete, stone, glass, wood, and tile.

Audio, video, lighting, motorized shades, security, HVAC (heating, ventilation, and air conditioning), and the pool/spa are all controlled from touchpanels located throughout the house. The lighting control system allows the owner to prepare the entire house for any situation. Whether it's an informal dinner with the family, a private supper for two, or a gala event for one hundred, lighting scenes can be activated to brighten the room, provide a soft glow, or highlight artwork. Distributed audio and video are also controlled with the Crestron system. CD and DVD systems display cover art on any of the system televisions. CDs and DVDs can be selected and viewed or listened to at any or all locations. AM/FM and satellite music channels can also be distributed to any of the twenty-four audio zones throughout the interior and exterior of the home.

Rigid and irregular building materials, a high level of fit and finish, and the need for simplicity all created a series of technical and aesthetic demands. The final result is an automation system that provides an appealing interface for family, friends, and associates.

Installer: Digitech Home Systems, Bellevue, Washington

Concert Quality

Sometimes a homeowner wants to tastefully integrate, yet be able to "show off" every type of electronics system imaginable: high-definition TV, plasma, multiple monitors, satellite, racks of equipment, dedicated theater, communication, cameras, lighting, and zoned control, not to mention zoned audio/video.

These homeowners, guitar collectors, had to reach a consensus as husband and wife when they undertook their electronic project. The man of the house wanted to show off all the gear, while his spouse wanted to hide it all away. With such an extensive equipment list, it took a little engineering to accommodate both. At this job, the cabinets were designed to make the stealth installation pop out at will. The equipment cabinet was designed as a closet behind French-style doors.

Some of the unique functions of this system include an AMX based control system that allows the house to be broken into twenty-four audio/video zones, thus allowing the family members to partake of a variety of pleasures without disturbing one another. Additionally, CD management is made easy by the Escient system that is tied into the audio/visual control system. CD Cover art is distributed through the video portion of the system while control and audio is available everywhere. Finding one title out of 1,500 takes mere seconds!

The Day Finger Pickers Took
Over The World
Chet Atkins
& Tommy
FINGER

Unique Installations

(Now You See It, Now You Don't)

Today's installers can make a TV screen pop up or a film screen unroll where you'd least expect it – from the foot of a bed, from behind a framed oil painting, or even over a bright sunny window. They can curve a screen to make it conform to a corner and surprise you with a wall of sound, artfully concealed. Rooms can be transformed in the wink of a button, and the film soon rolling. In fact, with electronic home engineering, you might even order the switch-over from your touch pad controls in the kitchen, or the bathroom, or,...

Pooled Resources

Here is a case where innovative ideas and implemented design made a huge splash!

These homeowners wanted to construct an addition onto their existing estate and to include a media system area in it. The proposed "addition" was half the size of the existing 10,000 square foot home, and the media system would be located in a 5,000 square foot indoor pool/recreation area.

The homeowners wanted to be able to relax in the pool or Jacuzzi and watch sports, television, or movies in this "multi-purpose" space. Lots of large windows allow natural light into the room and were left free of curtains and blinds since the owners decided they'd mainly use the media system in the evenings. The area also includes a large bar, salt water aquarium, and a unique Harley Davidson bike display area. The owners also wanted to be able to see the media system from the pool, Jacuzzi, sitting area, and bar. The challenge was being able to provide a clear view of the Harley display when the media system was not in use, so a 100-inch motorized screen was incorporated into a soffit above one of the glass walls. The architect and designer warned that all of the measurements for the A/V gear and sight lines to the screen had to be *exact*. Everything was neat as a pin.

An AMX Viewpoint control system ties in the whole house audio and lighting control systems. For a bright, vivid picture, the owner selected the Sony VPL11HT LCD projector and a Sony/B&W surround sound system all wired back to a Middle Atlantic AX-S Rack System housed in the Harley display area!

Installer: Electronic Design Group, Inc., Piscataway, NJ

MILLENNIUM
CRUISER

Photography by John P. Hamel

Creature Feature

This exotically decorated theater/living room combines the elegance of ornate cabinetry with the flair of high technology. Not to mention, it provides a nice place to lounge and enjoy company! Two gas-powered fireplaces flank the entertainment unit. Window blinds lower to darken the room. When the cabinet doors slide open, a 40-inch television is revealed. Six shelves above the set can house DVDs or CDs neatly out of sight. When movie time rolls around, a single button on the programmable touchpanel exposes a Vidikron projector in the base of the custom coffee table, and a 96-inch projection screen drops out of the woodwork, completely covering the cabinet space. Surround sound speakers are cleverly built into the wooden columns flanking the video center.

Installer: Advanced Residential Systems, Allentown, Pennsylvania

Photography by John P. Hamel

Photography by John P. Hamel

Photography by John P. Hamel

Photography by John P. Hamel

Artful Deception

These homeowners wanted to retrofit their formal living room to completely conceal every aspect of the audio/video technology, which includes a 61-inch plasma television, an equipment rack, and a surround flush-mounted speaker system with four subwoofers.

From a design perspective, there was no room for the equipment in the small, very symmetrical living room. To address this, a closet was constructed on the outdoor patio deck. It is hermetically sealed with rear access to the components, and the area was finished to match the outside of the home.

A white solar shade (2-feet wide x 7-feet tall) to match the interior walls is automatically deployed to hide the rack and all components when the system is not in use. When the system is operating, the screen is up; it then moves down to hide the gear when the system is off. An art form called "Giclee" (pronounced zhee-clay), was used to digitally reproduce an original painting, printed onto canvas. This limited edition fine art print of an original Steve Bloom, entitled *Champagne*, rolls up into the top of the picture frame using a roller mechanism and a contact closure. When the plasma is on, the art goes up; when the plasma is off, it is hidden by the artwork. The success of this project launched a new enterprise for the installers, "VisionArt," now a division of Solar Shading Systems. The company licenses original art works and creates these roll-up motorized retracting canvases to conceal plasma televisions.

Five of the nine flush-mount speakers are in the front wall. The entire wall was been fitted in light tone wall upholstery to completely conceal these speakers. The fiberglass padding (cut out where the speakers are located) helps address the acoustics of the room. The two rear channel speakers, as well as two additional subwoofers, are in-wall to finish an exemplary stealth and cosmetically pleasing installation.

Installer: Genesis Audio and Video, Irvine, California

Photography by Barbara White

Photography by Barbara White

Photography by Barbara White

And the Winner is . . .

The design concept for this room was an area where the homeowners could host an Oscar party, and which would serve at the same time as a cozy, homey living area for casually entertaining friends and family. The space in this elegant media room was maximized by the flexible arrangement of furniture. Tall sweeping draperies, plush fabrics, and rich colors provide the panache of English style. The interior designer, Barbara McLane, was able to create a comfortable, stylish environment, spiced up with the sophisticated use of furniture and accessories, such as a custom-built ottoman that cleverly houses the Runco video projector.

One touch of a remote control allows the room occupants to enjoy either a movie screen or television. The homeowners' favorite piece of art, hanging above the fireplace, can be transformed instantaneously into the large plasma television through the magic of VisionArt. The custom designed bookcases house the electronic equipment and provide additional media storage. For refreshment purposes, the homeowners included a lavishly decorated home bar for cocktails and late night snacks.

Installer: Genesis Audio and Video, Irvine, California

Panel of Expertise

This system was installed in an existing room originally configured as a mahogany paneled library. The room presented several difficult design challenges to the staff of Modern Home Systems: the ceiling is fourteen feet high; the solid wood paneling had to remain; a large bay window allowed in a huge amount of natural light; and no wiring or provisions for electronics had been made.

The system is based on the large Runco 9-inch CRT that lowers six feet from the ceiling into the proper position for viewing on a custom projector lift mechanism. The screen rises into position from a custom credenza that has a trapdoor hidden in the top. The credenza also houses all of the system components and speakers. The entire ceiling of the room was treated with several types of acoustic treatment materials to vastly reduce the reverb time of the room. The rear surround speakers were retrofitted into the paneling and custom painted to match the wood finish. The owners opted for a comfortable sectional seating arrangement in the front of the room. A riser with twelve custom finished theater seats was installed behind the primary seating.

The house features music in all of the main rooms as well as extensive sound in the outdoor entertaining areas. The audio, video, and lighting systems are controlled with Crestron wireless touchpanels located in the theater, family room, and master suite.

Installer: Modern Home Systems, San Diego, California

Disappearing Windows

A creatively designed theater allows the screen to drop down in front of the windows, blocking out unwanted light but allowing an attractive view when not in use. The speakers are concealed within the window bench below.

Audio Visions, Inc., Lake Forest, California

AMX

For Sports Junkies

For the avid sports fan, this unique and creative installation provides not only an excellent viewing environment for the game, it allows one to become part of it as well! In an adjacent basketball court, the kids can shoot hoops and play while dad kicks back and watches his favorite team in action.

Installer: Electronics Design Group, Inc., Piscataway, New Jersey

Entertainment Afoot

The foot of this custom-made bed consists of three panels in an olive green leather, embedded in a wooden frame. The central leather panel, which is operated by remote control, rises up to reveal a 20-inch flat-screen television.

Installer: David Vincent Design, Carmel, California

Photography by Tom Littlefield

Photography by Tom Littlefield

Complete Concealment

A family with a home overlooking Puget Sound in Washington wanted to upgrade their family room system while minimizing visual impact on the room itself. A visit to the home revealed that a projector and motorized screen would not work in this application since the projector would intrude into the adjacent kitchen and library areas.

A 50-inch plasma TV with a motorized lift was the preferred solution. The additional brightness of the plasma TV would also help minimize the effect of ambient light in the room. With the assistance of the room's original contractor, a cardboard mock-up of a 50-inch plasma was hung in front of the fireplace for the owner's approval. The homeowner's first response was "Do they make bigger plasmas?" A 61-inch plasma was soon proposed and the project went on to the next challenge.

Audio quality was of high importance but the windows surrounding the fireplace prohibited in-wall speakers. A pair of high-quality main speakers, custom painted to match the adjacent casework, were chosen to provide the main left and right channels. The center speaker was built into the TV drop-down box. Software was written to provide two-channel audio for when the TV is not in use. The rear speakers were replaced with custom painted speakers that, although bigger than the original cubes, actually blend better with the interior of the room.

When not in use, the casework containing the TV and center channel takes up a small section of the closet above this room. The existing trim piece was spliced and attached to the bottom of the TV/speaker box, making the retracted display nearly impossible to see.

The family's view of Puget Sound is now accompanied not only by the warmth of a fire, but also by their music of choice. When there's a game on or they are ready to watch a movie, a simple touch of a button lowers the plasma and the room becomes a full-featured media room.

Installer: Audio Video Resource, Mount Vernon, Washington

Big Screen Bedroom

A roll-down screen turns boudoir into big-screen entertainment via touchpanel control. The equipment is concealed behind the closet door, flush-rack mounted.

Installer: Genesis Audio and Video, Irvine, California

Photography by Anthony Gomez

Photography by Tom Littlefield

Pop-Up Videos

This unique and well-hidden plasma screen console was designed for a family living in a very upscale area on California's central coast. Traditional, elegant décor does not have to be sacrificed for high technology! The 50-inch plasma display is completely encased in the custom cabinetry (masterfully designed and built by Tom Littlefield), which beautifully accents the artwork and other home furnishings. When it's time to watch TV, the "on" button on a custom-programmed, touch-screen control system (not pictured) deploys a lift in the cabinet, which reveals the television.

Installer: David Vincent Design/Jim Zoyiopoulos, Carmel, California

Photography by Tom Littlefield

Disappearing Doors

This home theater is located on the lower level of a home on Coronado Island, an exclusive beach suburb of San Diego.

The room, with windows that open out to a beach level patio, presented a number of design challenges. The projection screen had to be oriented toward the windows and the view in order to work with the furniture layout and traffic pattern of the room. A custom soffit was designed above the glass panels to contain the flush-mounted, in-wall speakers, the custom electric film screen, and the retractable blackout shade. The 8-foot video projector was recessed into a custom pocket in the ceiling to minimize the distance the unit would hang from the ceiling.

The theater system, living room system, and house audio systems are located in custom racks in pear-wood cabinets flanking the fireplace. The cabinets also house a 32-inch high-definition plasma display and two powered subwoofers.

All of the equipment, including the theater, house audio, lighting, and shades, is controlled via wireless touchpanels, shown on the end table.

Installer: Modern Home Systems, San Diego, California

Picture Window

It's the best view in the house! And it won a CEDIA award for Best Hidden Installation.

This room had been preserved by the homeowner as a peaceful reading and relaxing room. Located high above the garage, it offers beautiful views from this lakefront home. When the family decided to add a media room, numerous options for its location were discussed. The basement was an option, but the owner felt it would be too isolated. The family room also came up, but since it was part of the kitchen and main space of the house, the owner felt it would be too distracting. Then the elegant lake view room was discussed, and an idea arose: preserve the room as a comfortable and functioning reading and relaxing room with lake views, but *transform* the space into a hidden technology room. Since the family wanted to keep the room's orientation towards the lake view, the idea was to design millwork around the windows to hide the screen and loudspeakers, creating a window seat effect.

They loved the idea when they saw it on paper! An interior designer immediately brought the room to life. Convenience features were added, including having the furniture on casters to allow easy moving and configuration of the seating. Within minutes, the seating can move from cozy entertaining style to theater style! The soft warm colors add to the peacefulness of this lake view room. The designer also helped keep everyone in check with regard to location of equipment. All of the equipment except sources that require loading were kept out of the room. The Middle Atlantic AXS rack is located in the adjacent room housing amplifiers, sound processing, control and switching components.

The subwoofers were hidden, and L/R channels are on each side of the screen with the center channel below. The screen is hidden in the bridge across the top of the window, while the projector is mounted on a lift concealed in the attic room above. The surround channels are all in-wall design. In order to make it more practical to use the room on weekends, motorized blackout shades were added to bring the room to darkness during the daytime. One motorized shade drops down behind the screen and three others cover all windows. All motorization and equipment operation is controlled by AMX.

Additional treatments include the ASC wall Damp System. The intention was to mask the sound transfer to other parts of the house. This consists of resilient channel construction (to keep the vibration off the wood frame) and dampening materials. Two layers of 5/8" rock complete the construction. The last part is a 1" thick solid-core door. Although this room is above the garage, it is off the main living area of the home and the owner felt it was important to mask as much sound as possible.

The homeowner has said several times that when entertaining new guests this is the last room they show. As guests enter they see a tranquil, peaceful room that flows with the rest of this warm inviting home. Then, with a simple button press on the AMX panel, the room comes to life – a jaw-dropping surprise every time!

Installer: Electronics Design Group, Inc., Piscataway, New Jersey

Hidden Surprise

These homeowners were delighted when the option to modify and clean up the unsightly mess of their existing home theater was presented. This was accomplished by mounting a 50-inch plasma monitor (with pull-out mount for optimum on-axis viewing from anywhere in the seating area), flush mounting the LCRs, locating the powered subwoofer and electronic components in the base of an existing piece of cabinetry, and placing a pair of small white bookshelf speakers, on adjustable mounts, amongst the decorative ceiling beams. Finally, to make operation of the system painless, a Panja color viewpoint package was added. An interior designer created beautiful Oriental panels that hide the monitor and speakers when not in use.

Installer: Integrated Media Systems, Sterling, Virginia

Room Transformation

This homeowner, a prominent attorney, and his wife have three grown children and a couple of young grandchildren. They wanted to be able to use their family room as a family gathering place and a place to entertain guests, while also having it serve as a home theater complete with surround sound. In order to do this, the home theater and surround equipment had to be installed with minimal intrusions.

The bright room has one complete wall of French doors that allow too much light for daytime home theater viewing. One solution was to program the AMX touchpanel remote to control the storm shutters that cover the French doors. A second area of consideration was the height of the ceiling in mounting the projector. A scissor lift was installed to raise and lower the projector. Finally, the owner wanted the center channel speaker to be as concealed as possible. The architect stepped up and designed the fireplace mantel to accommodate the center channel, which was then covered with a decorative iron grill.

Installer: Media Design, Inc., Houston, Texas

Sent from Above

This project was the very first Zenith 60-inch high-definition plasma television installed in the United States. While installing this ultra-modern new appliance, the design team was also challenged to maintain the country charm of the home and its setting.

The new construction master bedroom suite, sized at approximately 24 x 20 feet, was built to take full advantage of a wonderful meadow view and to implement a state-of-the-art master bedroom home theater. Some of the design complexities included a semi-vaulted ceiling, an extremely large picture window facing the king-size, four-poster bed, no planned furniture for local A/V equipment, and a desire for complete invisibility when not in use. In addition, the owners asked that electronic controls include the gas-lit fireplace.

Installer: AVIA, Huntington Station, Long Island, New York

Media and Multi-Purpose Rooms

(It's All Fun and Games)

Besides the plasma, mega-big screens, these rooms are packed with lots of extras. Generally huge on space, they include everything from pool tables to home bars. Some include picture windows that open to walk-out decks or patios. Others bring the outdoors in, as in the case of a lake-size aquarium teeming with fish.

Stainless Reputation

These homeowners had a very specific vision for their state-of-the-art home theater and music system in their billiard room. The homeowner is a true audio/video enthusiast and was very involved in product selection and the overall look of the room. This was his favorite room in the house, and he really wanted it to rock when he was entertaining! To this end, a custom designed and built 15-inch subwoofer was fabricated to fit his cabinet, and dual 6-inch, two-way in-wall speakers for the front LCRs were used to increase output.

The use of metal and glass in this traditional Mediterranean style house creates a truly unique style. The billiard room, for example, features a dramatic stainless steel ceiling. There was no obvious place to locate a big screen TV in the room, but the fireplace was clearly the focal point of the room. The only answer, therefore, was to incorporate a large plasma display over the fireplace. With the assistance of a very talented cabinetmaker, a compelling and entertaining room was created.

Installer: Home Theater Experts, Wilsonville, Oregon

Sound Advice

This homeowner placed audio above video when designing a media room, given a taste for classic rock-and-roll played at appropriate (loud) volume. A standard rear projection television was considered sufficient. The large room was also designed to house a bar and billiard table. Space limitations were a concern with the pre-existing cabinet, which dictated bookshelf-size speakers concealed behind acoustically transparent cloth. Genelec speakers, self powered and virtually indestructible, were perfect for the dual roles of home theater dynamics and accurate music reproduction.

Installer: AudioVisions, Inc., Lake Forest, California

Combining Work and Pleasure

This installation shows that a combination room doesn't need to mean compromise. One of the walls of this office is large enough to accommodate the homeowners' desired 100-inch Stewart Filmscreen. To maximize the potential of the screen size, a Runco CRT projector with a matched Runco scaler was chosen. The retractable screen lifts into the soffit above a cherry bench that can be used when the office is not in theater mode. The room's speakers are Linn, with three matched speakers as theater fronts and center channel, and smaller bookshelf speakers as rear surround speakers. Two compact Velodyne subwoofers provide low-end frequencies. System components include the flagship Lexicon MC-12 preamp/processor, with Proceed power amps for both theater and two channel listening, a Pioneer Elite DVD player, Marantz VCR, and Hughes HDTV receiver.

The system is controlled with a Crestron SmarTouch touchscreen remote that was customized to the homeowners' needs and wishes.

Installer: Audio Advice, Raleigh, North Carolina

Having It All

Imagine a warm room overlooking a lake, bright sunny windows everywhere, THEN...a button on the touchpanel engages ten motorized blackout shades and starts the show! These homeowners were seeking a complete solutions provider. They wanted various technologies, i.e. broadband Internet and console gaming, high-definition TV, theater-size screen, motorized shades, and lighting control. The design team consisted of a young talented architectural interior designer, who provided floor plans, millwork details, and color and fabric selections, plus an artisan who considers himself more a fine millworker from days past than a cabinetmaker. The homeowners were so impressed with the approach, they couldn't imagine doing this project any other way.

Installer: Electronics Design Group, Inc., Piscataway, New Jersey

Quickly Reconfigured

The seed for this project was planted during a routine service call for a previously-installed alarm system. The homeowner stated that she wanted to buy a "big screen." One discussion led to another and thanks to the service technician being on top of his game, a contract was soon signed to install a lot more than the "big screen" the homeowner was in the market for. The project was very detail oriented and took some twenty months to complete.

The room is centered on a large screen projection system,THX sound, and automated lighting. Ease of operation was accomplished with a color touchscreen control system. Source components include DVD, HDTV, DBS, VHS, and a PVR. The décor is complemented by custom woodwork throughout, including numerous marble inlays. The viewing area walls are covered with acoustical panels. This room is the homeowners' favorite place in the house for entertaining guests, playing games, and watching their favorite sporting events and movies.

Installer: Audio Video One, Winston Salem, North Carolina

Ultimate Hangout

Adjacent to the kitchen, a home bar, pool table, and expansive 64-inch HDTV screen make for the ultimate media/rec room. The shiny chrome-like finish continues from the bar area to the lighting system in the room and even onto the pool table!

Installer: Genesis Audio and Video, Irvine, California

This homeowner wanted his large unfinished basement to be transformed into the ultimate getaway. This area was a raw structure, so there was a lot of planning and hard work involved in the project. The ceiling needed reinforcement to handle the projector. Plumbing, central vacuum, and electrical lines were moved to provide more width to the room. The cabinets were custom-made off site and ended up being the only delay the project encountered. Efficient project management made the job come together extremely well.

The room is centered on a large screen projection system, THX sound, and automated lighting. The screen rolls away to display a large tapestry. Ease of operation was accomplished with a color touchscreen control system. Source components include DVD, HDTV, DBS, VHS, and a PVR. The décor is based on a wine motif and features custom woodwork throughout, including numerous engravings. The owner finds this room perfect for entertaining guests and watching movies with the kids.

Installer: Audio Video One, Winston Salem, North Carolina

This multi-use room combines a taste of the old with the latest in A/V integration. The overall design visually and technically meets the homeowner's needs. The goal of this Microsoft X-Box executive, his wife, the head of a non-profit organization, and their child, was to create a media room that would incorporate 21st century technology into a 1950s décor without altering the design of the room. Although the homeowner works with and understands the technology field, he wanted the audio, video, and lighting control systems to provide simple-to-use interfaces.

The room is constructed using an open-beam architecture with a loft and large windows that provide a bright, open living area. Unfortunately, this design also created several challenges with space and placement limitations. The placement of the screen and projector was designed around an old-fashioned telephone booth located in the center of the room. After reviewing several options, the design team decided to install the Seleco projector in the phone booth and mount the 120-inch motorized Stewart Filmscreen behind one of the beams. To further enhance the 1950s décor, all speakers were installed in-wall with the exception of one. The center channel was cleverly positioned inside a 1950s style radio perched atop the fireplace mantle.

To provide the twenty-first century appeal that the homeowner was seeking, an Extron interface was installed in the floor of the main sitting area. It connects an X-Box, PC, camera, or any other A/V device to the local projector and screen. The images and audio can also be controlled by a Crestron video switcher, which feeds the two 27-inch TVs that reside in the media room loft. This allows the homeowner to display DVD, satellite, and music, as well as test video game consoles and networking in this separate A/V zone. Lighting throughout is controlled by a Lutron HWI system.

Installer: Digitech Home Systems, Bellevue, Washington

Fished Up Nice

When a homeowner asks for an integrated audio and video system, things are pretty straightforward. This homeowner wanted just that, except that he was also building a new wing on his house for his fish tank and media center, and the final equipment location would be in this wing. What the homeowner really wanted was for the equipment to be installed for use in the current house first, then moved to its final home when the new wing was completed. An AMX controlled distributed audio and video system supplied signals to two 50-inch plasmas, a 42-inch plasma, and LCD monitors in each of the children's bedrooms as well as HVAC, lighting, and draperies.

The transition plan required that the system be down for less than two days! To accommodate the future location of the head-end equipment, each run had an additional eighty-five to one-hundred feet of cable that had to be managed so there would be no extra splices when the equipment was moved to the new wing.

In order to provide entertainment during the down period, the study – located in the older part of the home – was detached from the distribution system to allow use of a 50-inch plasma, DSS Receiver, and DVD player.

The end result was well worth the effort. Upon entering the room, visitors are greeted by a plasma monitor, four leather recliners, a pool table, and a 1,700 gallon fish tank measuring ten feet wide, five feet across, and six feet deep. Already a fabulous room for entertaining, the true magic begins when the movie button on the control pad is pressed. The lights dim, the shades draw, a 100-inch perforated screen and a smaller masking screen drop to conceal the plasma, and a projector descends nine feet from the cathedral ceiling.

Installer: Media Systems, Boston, Massachusetts

Plasma by day, screen by night...this media center offers it all. The entertainment center is the focal point of an inviting family room that opens to dining and kitchen areas. It is a room where the morning or evening news can be absorbed while other activities go on in the background. And when everyone is settled in for the night, the atmosphere changes to a wonderful home theater!

The 42-inch plasma television built into the cabinetry is available for ordinary viewing, but when it's movie time, the projection screen descends from the bonnet of the wall unit. A small Runco DLP projector deploys down out of the ceiling and the lift mechanism and the projector all fit in between the 12-inch ceiling joists. A programmable touchpanel enables easy operation and switching between the two video formats. The superb surround sound audio system heightens the experience, with the left, center, right, and subwoofer speakers concealed below.

Installer: Genesis Audio and Video, Irvine, California

Photography by Anthony Gomez

Photography by Anthony Gomez

Photography by Anthony Gomez

DEDICATED HOME THEATERS

Today's hottest must-have is a room that has it all over the multiplex down the street, with private seating, plush décor, and all the amenities of home. A lot of these even include their own popcorn machines.

A brand new feature in homes today, home theaters (formerly owned only by a few Hollywood insiders) have no décor dictums. They are conceived and executed in the spirit of fun and entertainment, and many reflect this in their design. The following images explore theaters that recreate the golden age of the silver screen through art deco motifs and Hollywood ephemera. Others unleash fantasy through jungle scenes, desert Harem motifs, Roman statues, and more. For those more conservative in their interior design tastes, there are plenty of theaters that mesh with today's home decorating trends. All are comfortable, all pack plenty of technological "wow" into their media presentations. Enjoy!

VAULTED TO THE TOP

This home theater was designed complete with a lobby – where the bar, refreshments, lighted marquis, and most importantly, the equipment, are housed.

The interesting challenge of this project was that the homeowners wanted to turn an unused "vault" in their home into a private theater. Being a vault, the room had a fixed envelope with 18-inch thick steel reinforced concrete walls, floor, and ceiling, and a 2 1/2-foot thick steel door. The key was finding a way to integrate bass traps into the room without them visually intruding. Owens Corning provided an acoustic room system treatment. The JBL Synthesis digital equalizer allows someone to fine tune the room even more, and made it possible to flatten the frequency response down to 1/4 db across the entire sound universe.

The end result was a classy, comfortable environment with phenomenal sonic and visual impact that invites "The Willing Suspension of Disbelief" and really gets viewers involved with the movies. An AutoPatch Video Router enables the video touchscreen to preview any source, and with a separate scaler to convert interlaced NTSC to 480P for HDTV display on the touchscreen, it acts as a HDTV television so the viewers can be watching a movie on the large screen while simultaneously monitoring a football game, etc., on the video touchscreen.

Installer: Performance Imaging, Greenwich, Connecticut

POPCORN

Playroom Grows Up

Before this room was a theater, it served as an exercise room and play room for the grandkids. The room was enormous, almost twenty feet wide by nearly thirty-five feet long with cathedral ceilings and exposed cypress beams running the length of the room. It had to be totally remodeled to create the theater. New walls and a ceiling had to be installed, the windows had to be blacked out, and, in order to get the theater up to the second floor (which is really like a third floor because of how large the home is), a cherry picker had to be brought in to lift up the pieces. The equipment rack was cleverly installed behind custom wood columns that add theatrical architecture to the room.

Installer: That's Entertainment, Lafayette, Louisiana

Thea-terrific

Bursts of light from dozens of little round bulbs and tubes of brightly colored neon frame the Uptown theater entrance, presenting the name of the current feature just above the door. Authentic movie posters are displayed on either side of the home theater entrance. The twinkling starlit fiber-optic design for the ceiling of this beautiful theater project originated in a theater in Brooklyn, New York. The Egyptian figures flanking the majestic red curtain are copies of originals above the marquee of the fabulous Pantages Theater in Hollywood.

Design by: Theo Kalomirakis Theaters, New York City
Installer: Robert's Home Audio Video, Los Angeles

This theater was designed for a Raleigh, North Carolina couple who moved from Atlanta and wanted to recreate the look and feel of their former hometown's Fox Theater. Outside the theater proper is a lobby area where reproductions of the Georgia landmark's ticket booth, marquee, and red velvet curtains set the mood.

Inside the theater, the ceramic tile flooring and cultured rock walls gave the room the potential for very bright sound. The solution includes the use of draped fabric to recreate the plaster "tent" that characterizes the Atlanta theater and insulating around the Bowers Wilkins Nautilus 800 series speakers that are concealed in the walls.

A Lexicon MC-1 processor enables sound to be tailored to the room's characteristics with two Bryston THX amplifiers providing power. A Sharp DLP projector and Faroudja processor, in combination with a Stewart four-way masking screen, provide a high level of video performance. The small projector, concealed speakers, and the housing of the components in a nearby equipment closet create an open, movie-house feel. Control of the Fox Theater is achieved with a customized touchscreen remote, which additionally commands twelve zones of whole-house music and automation.

Installer: Audio Advice, Raleigh, North Carolina

Projected Increase

This homeowner wanted to achieve the largest picture, with the most amount of seats, yet still be able to keep the décor consistent with the rest of the luxurious home. A CRT projector was definitely the choice, however with the tiered seating and high ceilings this posed the largest problem of mounting the projector into the ceiling. The projector would have had to hang low and the homeowner was not willing to sacrifice the look of the room with a large box built around the machine. Keeping the owner's biggest wish of achieving the most amount of seating, the designers had to carefully design the room for a table type box that would house the projector, yet still keep the look of the room consistent. The homeowner also insisted on sitting in the center of the room, not off to the side. Eventually, eleven seats fit in this 17 x 23-foot room, with large walkways on each side as well as in the rear of the room.

In keeping with the décor, the ceiling had to be Victorian style. Originally, the thought was a tin ceiling, which obviously would have been an audio nightmare. So a wallpaper material was used instead, painted gold to give the look of a tin ceiling. The ceiling was then complemented with a custom backlit emblem, along with indirect cove lighting.

The homeowner was also very insistent with regard to his choice of materials. Irwin Seating took care of the seats. When it came to the wallboards, the owner was very budget conscious, as well as hands on. As a result, he decided to make his own panels with his own material, doing an excellent job.

Installer: Sights-n-Sounds, Long Island, New York

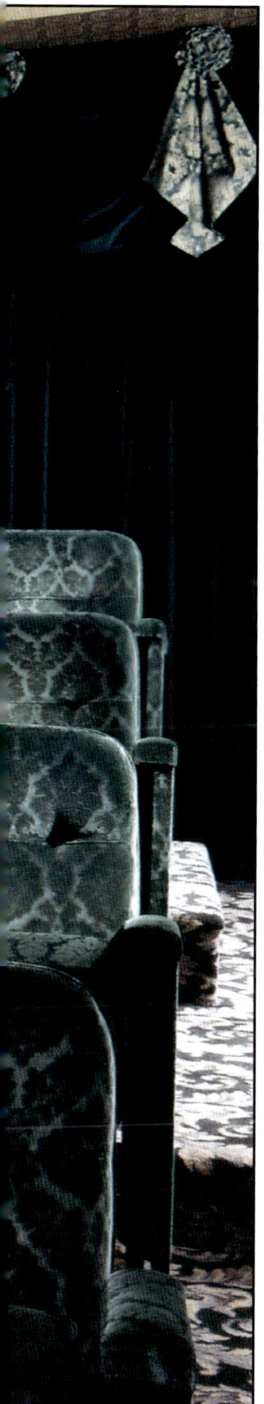

Complete with Popcorn

The most outstanding architectural feature of this home theater is the room's ability to envelope you and take you back to a time when theaters were lavish and ornate, full of palatial splendor and color. The most outstanding audio/video feature is the marriage of the latest control and surround technology with the most film-like video system available today. The owner was an avid moviegoer in her childhood city of Burbank, California. The room's architectural style was designed in conjunction with those memories and the feel of those theaters. The ornate wrought ironwork is mirrored throughout the home, including the front driveway gates. This room provides a fully functional theater experience, complete with a cinema bar that includes a fountain drink dispenser, popcorn, candy display, and all major appliances. There is a power room decorated in a whimsical fashion indicative of the era with famous movie quotes stenciled on the walls and ceiling.

The obstacles were mostly structural and acoustical in nature, regarding the placement of the Vision One projector, seating requirements, site line, and distance to the screen, while retaining the overall palatial concept. The ceiling was raised in the theater area from the existing 8-foot mark to the 12-foot peak vaulted partition. By vaulting the stairway landing ceiling to emulate the theater side, a more ornate ceiling was created. Sheetrock was moved up in between the rafters on the cinema bar side to create coffers. Drapery, enclosures at each dormer, and window openings were all designed as well, many of which had been acoustical considerations. Columns, wall treatments, and ceiling coffers were added to aid in sonic room control.

Installer: Nicholson's Hi-Fi, Nashville, Tennessee

Art Deco style is enhanced with vintage posters and unique side tables made from old film reels to create a classic Hollywood atmosphere.

Installer: Audio Video One, Winston Salem, North Carolina

Rudolph Valentino
GEORGE MELFORD'S PRODUCTION
"THE SHEIK"
AGNES AYRES and RUDOLPH VALENTINO
It's a Paramount Picture
HAL ROACH
STAN LAUREL AND
OLIVER HARDY
LIBERTY
A Metro Goldwyn Mayer Picture

What movie theater is complete without a fully stocked candy counter and arcade? This true-to-life lobby comes complete with an amphibious ticket window and a friendly butler waiting to greet you with some samples of today's treats. Some retro video games and an antique popcorn machine add to the ambiance, and an authentic marquis displays what's featured in the theater.

Inside the theater, old movie posters adorn the walls and three rows of genuine movie theater seats accommodate twelve viewers. A Vidikron projector mounted to the ceiling and a display router are controlled by a wireless touchpanel to bring the large screen to life!

Installer: Advanced Residential Systems, Allentown, Pennsylvania

Photography by John P. Hamel

Photography by John P. Hamel

Photography by John P. Hamel

Photography by John P. Hamel

New Construction Retrofit

These homeowners purchased a unit in a "high-end" townhouse development that was under construction. Their desires included a traditional yet not ornate environment, and outstanding video quality. Because the owner had a background as a professional musician, a major requirement was the ability of the sound system to provide theater excitement and true, life-like music reproduction. This room is used more than fifty percent of the time for musical enjoyment. The last and most often heard request was that the system be easy to operate.

Oddly, that was the easy part. The challenge started when the builder did not allow any "jewelry" to be added until after the closing. Therefore, work could not start until the closing of the property. Thus, a project that had been designed as new construction became a retrofit, and now included removing walls and doors that were part of the builder's original spec, but not the theater's. This truly became the challenge and taxed available resources, yet added to the final satisfaction of the homeowner and of the installer.

The biggest satisfaction was hearing the homeowner say at the close of construction, "I'm very proud of it... My theater sounds like a professional studio."

Installer: Electronics Design Group, Inc., Piscataway, New Jersey

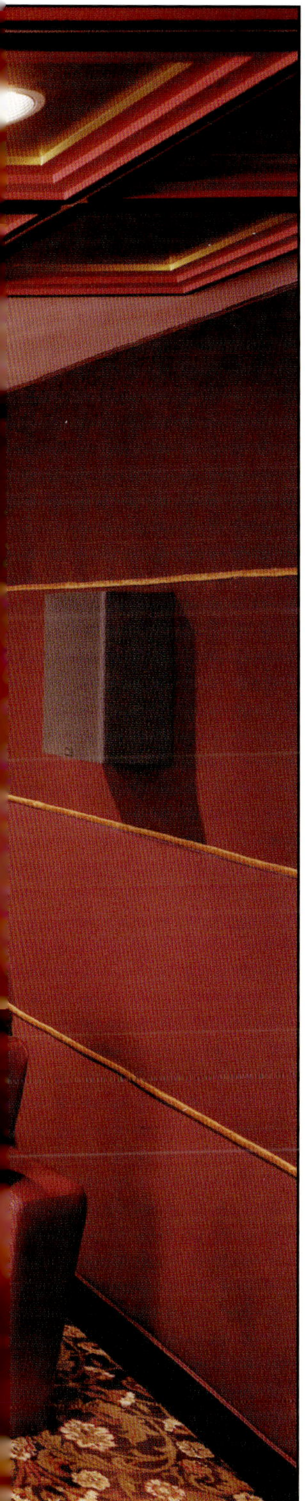

© 2003 Electronics Design Group, Inc.

© 2003 Electronics Design Group, Inc.

© 2003 Electronics Design Group, Inc.

Mini Radio City

The concentric arches in the auditorium were inspired by similar structure at Radio City Music Hall. The vents create architectural interest and serve two purposes: to allow light to pass through, and to provide HVAC, which was the original function of those found at Radio City. Plush, velvety red seats give an authentic movie theater feeling, while snapshots of Hollywood's finest adorn the walls in black and white.

Design by: Theo Kalomirakis Theaters, New York
Installer: Audio Command Systems, Westbury, New York

Absolutely Acoustic

Simple in design, the trimwork and upholstered walls lend themselves to perfect acoustics. These theater buffs were going for the ultimate cinema experience at home, starting from the entrance. The project features a Runco DTV-930 projector, a Runco sealer, Parasound and Bryston electronics, custom made Green Mountain Audio speakers and subwoofers, and an AMX Viewpoint control system.

Installer: Connected Technologies, Colorado Springs, Colorado

Photography by Cheryl Ungar

Photography by Cheryl Ungar

Photography by Cheryl Ungar

A Colorado radiologist had this theater created with much of his own gear spread between two different rooms. There was no attempt to hide the equipment. In fact, this room flaunts it!

A new room was designed in the unfinished portion of his home, and he integrated his home theater equipment and audiophile listening equipment into it. The audiophile system features Infinity IRS Beta speakers, with AR and Threshold electronics. The home theater system utilizes a Runco projector, Snell and Triad THX speakers, a Theta processor, and Adcom amplifiers. A Niles Audio Intellicontrol makes the whole thing manageable. The owner found the Charlie Chaplin statue while on vacation in Florida and wanted to create a place for it. Charlie now greets all guests as they enter the theater!

Installer: Connected Technologies, Colorado Springs, Colorado

Photography by David Beightol

Photography by David Beightol

Photography by David Beightol

Photography by David Beightol

Bogey on Course

The Bogey Theater incorporates whimsy and pun in the entrance to a luxurious home theater in a golf course community. A custom designed ticket booth and life-size moving and talking replica of Humphrey Bogart greets guests.

Inside it's all business, with acoustic paneling, a double rack of equipment, and an enormous projection screen. Ten custom leather theater chairs recline, and were arranged in tiered pairs so that every seat has a perfect view. The master of the house has the touch-operated panel to control lighting and the screen at his fingertips, right between the two front-row seats.

The projector is encased in a cherry wood cowling that matches the room's paneling and grid-patterned ceiling. The woodwork and slate accents work with the rest of the home's décor. An acoustically transparent screen conceals the audio equipment and fabric-covered panels hide the in wall speakers.

Installer: Room Service Home Theater Specialists, Portland, Oregon

STEWART
REED

Reel Solid

This theater was created for husband and wife Olympic figure skaters. It features a Runco DTV-850 projector, Yamaha processing, and Atlantic Technology speakers. The reel-adorned carpet spices up the room.

Installer: Connected Technologies, Colorado Springs, Colorado

Photography by Ron Johnson

This theater project, spanning only 250 square feet, turned an otherwise ordinary basement into a hi-tech, fun-filled entertainment center, complete with maple dance floor and fully stocked bar. The room has curtains for walls, sacrificing only minor acoustical elements and allowing theater occupants to close off the whole room for a theater feeling or only part of the room so there is access to the bar. An 82-inch Stewart Screen hides behind the elegant purple curtains and the projector is creatively concealed within an oversized, decorative film reel hanging from the ceiling. A black Plexiglas plate is blanketed with hundreds of fiber-optic points of light.

The Art Deco-style modular theater chairs, in beige and purple faux suede, are big recliners for a more comfortable and homey theater experience.

Installer: Howell & Associates, Inc., Burlington, Ontario, Canada

Bedroom Conversion

A fifth bedroom conversion in a waterfront Florida residence became the household movie theater, complete with additional video projection, 7.1 digital surround sound, and a 120-inch movie image. Sumptuous full motion recliner seating on two tiers was installed, along with acoustical paneling, accent lighting, and authentic motorized movie theater screen curtain. Outside the theater is an entrance marquee.

Installer: First Impressions Theme Theatres, Miami, Florida

Photographers2, Inc./Jonathan Rachline

Reel Ape-peal

Up to eight can lounge comfortably in this climate-controlled, primate-adorned home theater. Ten speakers issue impeccable surround sound, while a 6-foot wide, 16 x 9 aspect-ratio screen provides a great display for DVDs, high-definition television, or even video games! A brand new, cutting-edge DLP projector provides the images for movie time, and any of the owner's six hundred CDs can be called up at the touch of a Crestron remote control.

Installer: Howell & Associates, Inc., Burlington, Ontario, Canada

One-upmanship in the Making

On December 26th, this homeowner left a message on the theater company's voice mail telling them that his wife had purchased a shiny red Ferrari 360 for herself as a Christmas gift and it was sitting mysteriously in their driveway on Christmas morning. In his message he stated the following: "My wife just comp'd herself a Ferrari and I have raised her 'a home theater.'" His only requirements were that the theater cost more than the Ferrari Modena and be painted red, and that the final statement be addressed directly to his wife for his amusement.

The project was simple and it went smoothly. The homeowner wanted a "hot-seat" section where stimulus seekers could sit closer to the screen and front speakers, and he requested relaxed viewing further back in the theater. Meridian made it easy to program the front and rear speakers so that they catered to the front two theater seats. Meridian's sound control advantage combined with the plush seating and tray ceilings neutralized any need for heavy acoustic work.

Easier than a five-speed gearshift is the Crestron touchpanel that summons the homeowner's source of choice and dims the lights accordingly. Within his budget we did our best to make this environment and system as gaudy and chic as his wife's Ferrari, but one that she would enjoy as well, long after the joke was over.

Installer: Atlanta Home Theater, Georgia

With an Eye on Baby

This room was intended to be a full-scale theme theater from the beginning. A soundproof room, stadium seating, and velvet curtain create the true theater feel. The owner wanted all of the equipment hidden to keep the emphasis on the room itself and not on the electronics. As a result, the only visible item is the AMX touchscreen control. The projector is neatly hidden in the rear soffit, and all the electronics are kept behind one of the acoustic wall panels. If you didn't know where they were you would never find them. To monitor the baby's room while they watch a movie, the homeowners use the touchscreen with panel video. In the event that the baby cries, audio detectors pick up the noise and automatically switch the image to the big screen to alert the parents.

Installer: Designer Audio, Naples, Florida
Builder: Gulfshore Homes

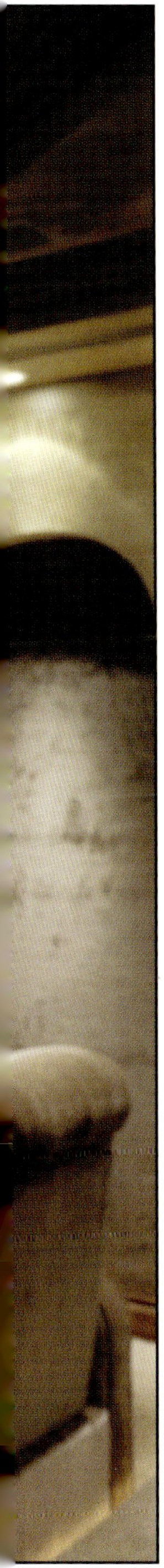

This beautiful theater won the award for Best Home Theater, Level IV $260,001-$380,000, in CEDIA's 2001 Electronic Lifestyles contest.

These homeowners had a traditional theater in mind, and a mural of cowboys riding off into the sunset was suggested for the rear wall to complement the rural Oregon location. The home is a 6,000 square foot stick frame addition to a log cabin, with a lot of rock and wood features. Acoustic Innovations was selected to assist with room design and to provide materials and installation of the décor. They provided color renderings, fabric samples, and their best artist to complete the mural. The original architectural plans, with two exposed beams and a short room length, made the room difficult to configure. The room was therefore lengthened and the beams were included in the ceiling décor by adding cross sections to form a grid, complementing the fiber optic ceiling panels and shooting stars.

The first objective for all electronic design is to make it easy to use. By using Crestron's CNMSXPro processor and TPS-4500 touchscreen, the homeowner can easily control the theater system, access sources from the house audio system, and control the Homeworks™ lighting system and Europlex security system, all without leaving the "best seat in the house."

The homeowner was specific about having high quality theater sound and exceptional video quality. After the room was reconfigured, it was necessary to upgrade the JBL Syn2-L7, the Runco DTV-1100 CRT projector, and VFC-4404 video processor to a JBL Syn1-L7 and a Runco VX-3C DLP projector that is now hidden from view. It projects through a projection glass window in a soffit in the back of the room, which also helps with fan noise. The Escient PowerPlay with the Crestron provides an easy user interface to access the DVD library, and completes the system with distinction.

Installer: Premier Home Theater, Bellevue, Washington

Photography by Leo Arfer

Equipment List
1 Cinepro Power Pro 30
1 Furman AR-1230
1 Furman PL-Pro
1 Crestron CNCS
4 Crestron CNIRP
1 Crestron CNMSX-PRO
1 Crestron TPS-4500
1 Crestron TPS-VID
1 Crestron CT-1000
1 Escient Power Play
1 JBL Synthesis Syn 1-L7
1 Marantz DV7000
1 Marantz MV-880
1 Middle Atlantic MRK Rack
1 Middle Atlantic Slim5 Rack
3 Middle Atlantic RSH-4a Rack Shelves
3 Middle Atlantic D-4 drawers
1 Middle Atlantic D-3 drawer
1 Proscan PSHD105
3 RF systems VS-B
1 Runco VX-3C projector & controller
1 Stewart Quad Electro-mask screen
Wireworld Wire & cables
1 Acoustic Innovations Theater room décor, including:
acoustical panels, mural, mouldings, curtains, fiber optic ceiling, proscenium, and shooting stars
10 CinemaTech Lonestar Theater Seats
1 CinemaTech Console
2 Premier Home Theater Custom game ports

Photography by Leo Arfer

Photography by Leo Arfer

Mayan Magic

Step into this theater and find yourself somewhere between the Great Andes and the plains of Patagonia deep in the Amazon jungle, home of the Mayan people. Set between the doorway of an abandoned temple and its altar of worship, this habitat is alive with the sounds of wild birds and animal calls heard beneath the waterfall. Step carefully through the plush tropics and gaze at the exquisite authentic markings made of yellow ocher, lime, hematite, charcoal, blood, and other pigments illustrating gods and relics of that time. Notice a purple haze descending from the starry sky over the indigenous landscape of fossilized ferns and flowers planted directly into the theater's floor. Observe the tiny lights illuminating the stone pathway and the massive columns marking the entry. The Mayan god of light was carved with precision into the altar between two flickering torches. The god of water and the god of corn are seen holding a DVD and remote in their hands.

Installer: Atlanta Home Theater, Georgia

Ancient and State-of-the-Art

A Greco-Roman theme in marble structure and trompe l'oeil finishes creates "awe" power for this home theater. Per the owner's request, the control of acoustics, lighting, window curtains and blinds, the video projector, and other home theater equipment can be carried out both manually and automatically.

At first sight, this room was not suitable for a home theater. An equipment closet was specially designed to tuck under a staircase, situated in such a way that it is possible to service all these sources of different signals without entering the closet. Within, one wall is used for home theater equipment (screen, front and central speakers, two LCD TV sets, and signal sources); the second wall is used for Crestron lighting and the curtain control equipment; and the third wall is occupied by audio equipment and shelves for CDs and DVDs. The whole set of equipment is controlled by a Crestron system and every room in the home has push-button wall panels, remote controls, or touchpanels.

Installer: En-Trade, Russia

Cozy Escape

Rich with warm brown and honeyed tones, this cozy theater is all about comfort. Curtains lower over transom windows to seal off natural light when the show gets under way and the star-riddled fiber-optic proscenium provides a dreamy, nocturnal atmosphere. The rich upholstery extends to the wall panels, where cloth adds to the acoustical qualities of the home venue.

Installer: AVL Pro, Inc., Naples, Florida

Fair Play

A Renaissance-festival atmosphere was created for a local doctor and his family. The theater features a Runco DLP projector, Atlantic Technology speakers, and a Niles Audio control system. The custom painting features a likeness of the family.

Installer: Connected Technologies, Colorado Springs, Colorado

Photography by Roger Wade

Starship Style

This theater was created in cooperation with the architect, builder, and interior designer during new construction. The goal of the owners was to create a theater that would simulate the bridge of a Star Trek Federation Starship.

For the theater's interior, curved rear soffits and a domed ceiling (with fiber optic stars) were incorporated into the design to create the illusion that the rectangular room was round. Powered Meridian speakers are hidden in the rear soffits and in the walls at the front of the room, further enhancing the Starship atmosphere. In synergy with the speakers, a digital Meridian surround processor and a CD/DVD player are the core of the system. Clark tactile transducers are installed in the seating to enhance the movie viewing experience. An Escient 200 DVD jukebox, Sony S-VHS VCR, and an HD DSS receiver are the other primary source components.

Primary video source material is provided by a Runco VX3C DLP projector, and displayed on a Stewart Filmscreen. Flanking the main screen are two Sony 20-inch direct-view televisions, providing the ability to follow more than one sports event at a time. A Crestron touchpanel controller operates the theater, with the custom keypad design simulating a template from the Enterprise.

Installer: Audio Advice, Raleigh, North Carolina

Box in Box Architecture

The founder of a very successful dot-com company, this homeowner is aware of and interested in all that is new, high tech, and cool. He wanted the best for his home theater project.

Residential code restrictions limited this retrofit project from being built out from the home, so the designers had to go down. Built on a hill, the house had some good crawl space areas. For months a crew dug by hand, as there was no way to get machinery into the space. Working with architect David McNiven of Oregon, who specializes in home theater design, the old post and beam foundation was moved to make way for a new span beam system. All the foundation walls were rebuilt as well to allow for the much taller space taking shape. The very large addition to the home allowed for an eleven-seat, three-row state-of-the-art theater, a lobby/bar area, a powder room, and a beautiful wine cellar.

Because the family room and kitchen areas were directly above, the family needed an area that was completely isolated. The new theater incorporated McNiven's box in box architecture: the vast, dug-out area for the theater was framed conventionally, then isolation couplers were applied to the studs, metal attachment strips run through the couplers, and drywall applied. The drywall never touches the studs, creating the first sound barrier. In addition, the theater footprint is framed one inch inside the first box. The inside box doesn't touch the outside box, creating the second sound barrier. Though McNiven allowed penetration through the outside box for things like wiring and mechanicals, nothing else was allowed to penetrate the outside box. The results are astounding.

The owner is putting the theater to good use. He quickly filled his first 200-disc DVD changer and had to add three more shortly after the installation.

Installer: Room Service Home Theater Specialists, Portland, Oregon

Sound Dedication

This beautiful home theater was designed to fit an existing room in the owners' 10,000 square foot rural estate. A "floating" mid-room riser was utilized to permit tiered seating while allowing adequate aisle space on the sides. This provided room for a leather reclining loveseat on the riser and three individual leather recliners in the front row.

The home theater was outfitted with a DLP projector, a complete THX Ultra EX speaker package, a THX surround receiver, THX room equalizer, DVD player, a C-band satellite high-definition receiver, and a programmable touchscreen remote. The remote was programmed to automate not only the media equipment but also the screen curtain control, and the room's lighting.

One wall of the theater was adjacent to the home's mechanical room. Acoustic treatments were utilized to keep the mechanical room noise out of the theater and a unique sound-proof dual door was designed for access to the mechanical room. The room's trim and acoustic treatments continue unbroken across the face of the door, rendering it "invisible."

Photography by Jim Wark

Photography by Jim Wark

Photography by Jim Wark

Movie Memorial

This homeowner is in the business of building bridges, and wanted to incorporate a famous bridge into the design of the room. The family felt that the Brooklyn Bridge and the New York City skyline would be an appropriate tribute to the victims of the September 11th tragedy.

The New York City mural was pre-painted on vinyl and later adhered to the theater walls. The skyline wraps around from the back of the room to the viewing area, right into a three-dimensional Brooklyn Bridge. Small steel rods were used to create suspension, and fiber optics were installed directly into the drywall before the ceiling was mounted.

Since entry into the media room would be at the back right side of the room, the owners did not want a rear seating riser to be visible. As a result, a step and landing were designed outside the double door entry. Once extended into the room, the landing became the main floor for the back row of seating and the bar area, with a step down to the first row.

The client's extensive DVD collection is easily accessible through the Kenwood Entre Entertainment Hub and the 401 disc DVD changer. DVD cover art, disc name, and genre are retrieved via the Internet. The music collection is stored on an Imerge Sound Server and can be accessed on the screen or the two-way wireless Crestron touchpanel.

Installer: Howell & Associates, Inc., Burlington, Ontario, Canada

Matrix Motif

This fully customized, flagship theater was inspired by the homeowner's request for a *Matrix* themed, dedicated home theater room. It won CEDIA's Electronic Lifestyles contest in 2002 for Best Home Theater, Level V – over $900,000.

Knowing that there would ultimately be a second, third, and possibly even a fourth *Matrix* movie, the homeowner wanted his theater to be thematically expansive. The interior designer, Lezlie Trujillo, rented the movie and watched it three times in one day, taking copious notes as to how to design the theater to be cosmetically appealing.

Control of the entire system is performed with the AMX 10.4 control panel. All aspects of the theater, as well as the house, can be controlled through this touchpanel. Cameras can be viewed, doors can be locked or unlocked, shades can be raised and lowered, windows can be opened and closed, and the spa can be turned on and off. The entire house lighting and security system is also integrated into this theater touchpanel.

Additional "creature comforts" include two rows of terraced leather seating for eight individuals, each with its own motorized-recline controls, plus three ultra-suede conical, swivel chairs for the bar area. The room also incorporates acoustic isolation, a floating floor, and an ultra-quiet dedicated air conditioning system.

The designer set out to make the homeowner's interior design requests a reality. Columns similar to those in that fantastic scene where Keanu Reeves and Carrie-Anne Moss shoot it out conceal the main and side speakers. The center channel speaker is located under the screen. The Artemide sconces echo the film's "breeding colony" concept and the stainless-steel door to the equipment room is similar to Morpheus' submarine door. The ceiling detail represents a thunderous sky and the center cloud conceals the projector when not in use.

Installer: Genesis Audio and Video, Irvine, California

Equipment List

1 Runco DTV-1101 CRT Projector
1 Faroudja DVP-5000 Video Processor
1 SVS Projector Lift
1 Stewart 4-Way Masking Diagonal Screen
1 Escient Power Play Media Jukebox
2 Escient MC-201D Changers
1 JVC VHS/DV VCR
1 RCA DTC-100 HD DSS Receiver
1 Sony-DVP-9000ES DVD Player
1 Cinepro Power Sequencer
1 Cinepro Power Balancer
1 Lexicon MC-1 Audio/Video Processor
7 Linn Klimax Solo Amplifiers
1 Linn Klout Power Amplifier
2 Revel Studios Speakers
1 Revel Voice Center Speaker
2 Revel S-30 (Side Speakers)
2 Revel Embrace (Rear Speakers)
3 Revel B-15 Powered Subwoofers
2 Middle Atlantic Equipment Racks
1 AMX AXT-CV-10 Touch-panels
1 AMX HUB Control Hub
1 AMX CF-10 Card-frame
1 AMX VSS-8
1 AMX PLC-AMX
1 AMX MCU
1 AMX PLC IROUT
1 AMX PLC IN-7
2 AMX PLC RL-8
2 AMX PLC SER
14 AMX PLL-MLC
2 AMX UPC-20
1 Statnet PCC-300 Thermostat Controllers
8 Cinematech Electric Chairs
2 Acoustic Innovations Fiber Optic Stars System

Photography by Anthony Gomez

Photography by Anthony Gomez

Photography by Anthony Gorrez

Perfect Room

A marriage between an audiophile and a movie lover resulted in the birth of this theater. It was built new, as part of an "entertainment wing in a home," and the project managers found that their job was easy because they helped to make sure the room was built right for sound. After all, the room itself is the biggest, most expensive piece of gear in a home theater system, and when it's done right, very little tweaking should be necessary. In this case, the wall construction included two 2-inch layers of insulation separated by two inches of air, and the perimeter walls sandwiched a soundboard between two layers of drywall. Moreover, solid, handmade mahogany doors add to the nearly soundproof environment. The coffered ceiling serves as a defractor for sound, along with looking stylish. The source components are separated from the rack of electronics so it's easier for the homeowner to change sources. This is a seven-channel, fully electrostatic home theater, using Martin Login speakers. It features a New Orleans theme, with statues in a cubbyhole at the back of the room representing an authentic Cajun jazz band.

Installer: Genesis Audio and Video, Irvine, California

Forefront of Elegance

For this project, an unused room at the front of the home was transformed into an Acoustic Innovations theater space. The homeowners asked for maximum seating, resulting in the slightly off-center configuration in the 19 x 22-foot space. Now eleven can watch in comfort. The black and gold décor reflects "California elegant" style, in accord with the homeowner's tastes.

Installer: Genesis Audio and Video, Irvine, California

Photography by Anthony Gomez

Photography by Anthony Gomez

Drive-in Theater

This eight-seat miniature home theater used to park one car. A garage remodel, it is now a pleasingly packed 10 x 16-foot space, stocked with a great equipment lineup including eight-speaker surround sound, a Stewart screen, Runco Super DTV-933 projector and line tripler, and a Lexicon processor.

Installer: Genesis Audio and Video, Irvine, California

Photography by Anthony Gomez

Photography by Anthony Gomez

Fading Beauty

These homeowners invested $230,000 on a bonus room in their home, creating an opulent theater. The striking red and black theater had to be completed in just two short months. The end result is stunning. At the touch of a button, the curtains begin to draw back and the lights dim gradually over a thirty second period, preparing the view for the feature presentation. An attached bar area allows someone to pour a drink or get popcorn without leaving the action or changing the lighting in the theater.

Installer: Genesis Audio and Video, Irvine, California

Photography by Anthony Gomez

Photography by Anthony Gomez

Photography by Anthony Gomez

Custom Contemporary

This auditorium was designed to match the contemporary architecture of the home, right down to the custom designed seats. The mahogany-and-granite bar at the rear of the room provides additional seating, and houses the Apollo's equipment. The walls are glazed, rough-textured plaster, varied by zebrawood paneling arranged in a checkerboard pattern. The wall's sculptural elements conceal the surround-sound speakers, and the floating ceiling conceals the projector. Upholstered acoustical panels to either side of the screen soften the emphasis on hard surfaces elsewhere in the room, while absorbing sound reflections to enhance acoustics.

Design by: Theo Kalomirakis Theaters, New York City
Installer: Genesis Audio and Video, Irvine, California

Olympic Glory

Inspired by the homeowner's visit to Teatro Olimpico in Vicenza, Italy, this design team was commissioned to create a unique private cinema approach with an indoor/outdoor amphitheater flavor. Ancient stone and marble architectural elements were replicated using mahogany and textiles. This new "digital age" Teatro Olimpico features hand-illustrated fresco artwork as a backdrop, carefully airbrushed upon transparent scrim that covers integral bass-traps and sound diffusers. The sidewalls, void of wood elements, were clad with acoustical scrim over 2-inch pure cotton sound-dampening insulation. The coffered ceiling panels contain 4-inch absorbers and diffusers strategically placed for optimum audio playback. A multi-dimensional "dusk to dawn" fiber optic Starlite® ceiling was added to marry the wall scenery up and onto the ceiling design.

Located on the second floor of a palatial Miami-area residence, the theater's biggest challenge for the design team was sound containment and vibration control generated by the "overkill" multi-speaker digital sound system. As a stick-framed construction, with living space right, left, and below, a room within a room build-out system was developed to include floating risers and sub-floor, decoupled wall assemblies, and a ceiling hung on deflection springs. Complete with a sound seal door and one room HVAC system, the theater shell, before acoustical treatments, was a guaranteed top performer.

A projector bunker was built in the rear of the theater, becoming a soffit in the adjoining bedroom.

Installer: First Impressions Theme Theatres, Miami, Florida

Photographers2, Inc./Jonathan Rachline

Photographers2, Inc./Jonathan Rachline

Photographers2, Inc./Jonathan Rachline

A Guest Room, Too

This private in-home movie/video theater based in the penthouse of an opulent Miami Beach residence serves a dual purpose. It is a private cinema for the owner's enjoyment of DVD movies and a music listening room for CDs, as well as a guest room on a need-be basis for grandchildren. The front row of this theater's CineLounger® seating becomes a sleep sofa. Careful construction details were implemented for a totally soundproof environment so as not to wake up the neighbors living above or below in this condominium residence during action/adventure movies or videos. The entire theater was built in First Impression Theme Theaters' Miami-based factory, transported to the project site and installed by the company's capable craftsmen. The theater features a 60-inch movie image, high-end THX sound system, and Crestron automation.

Installer: First Impressions Theme Theatres, Miami, Florida

Motorcycle Magic

The Harley Theater, a superbly detailed and designed entertainment room, was created by Jeffrey Smith of First Impressions for this collector of Harley Davidson motorcycles. The proscenium columns echo Harley gas tanks as does the aisle detail on the CineHarlee™ loungers done in red and black leather. Harley gas caps are used to cover the cup holders when not in use. A streamlined, Harley-like gas tank acts as the shroud for the Sony CRT projector. A Faroudja scan converter provides a pristine image on the 100-inch Stewart screen. An ADA processor and amplification package delivers the signal to Miller and Kriesel 5000 THX speakers, all discreetly hidden below the screen and in custom speaker sconces. An AMX Color Video touchscreen controls all automation in the room, including the motorized Austrian style drapes and LiteTouch Lighting control system. The owner can even view who is at the front door or gate by using the touchpanel. The theme continues in the adjacent bar and rec room.

Courtesy of Monaco Enterprises, Inc., Spokane, Washington
Installer: First Impression Theme Theatres, Miami, Florida

Photographers2, Inc./Jonathan Rachline

Photographers2, Inc./Jonathan Rachline

Harem Stare-um

How do some keep the excitement alive in their daily existence? Many dream of visiting far off places, but can't find the time to break away from everyday life. Wouldn't it be wonderful if an exotic escape were only downstairs? This oasis is a hidden gem in the basement. A theater built for two!

By incorporating an Arabian harem theme, these homeowners created a place little known to guests and friends. It is their secret place where together they can enjoy each other's company and escape to the magic of the movies.

Upon entering the room, you are immediately transported to an enchanted land. Looking at the stunning surroundings, it is hard to believe that high tech equipment is hidden in this room. An elegant tile border surrounds a midnight blue carpet, upon which rests a precious family heirloom rug. Hidden behind delicate carvings on the wall, the speakers and woofers are completely out of sight. Every detail was personalized, making this the family's favorite place. From the thematic sconces to the multi-colored leather trim around the room and the luxurious red velvet curtains with gold rope tiebacks, no detail has been overlooked. As the lights dim the true intent of the room is put to test!

This intimate theater displays sound and video quality surpassing many professional screening rooms! The 130-inch perforated aspect-ratio masking screen displays images provided by a performance-caliber Digital Projection DLP projector. A Lexicon digital surround processor ensures sound is properly distributed throughout each of the recording-studio caliber Genelec cinema speakers. During intense action, bass shakers beneath each seat create a vibration that rocks the viewers in their seats. Combined, the elements of design and electronic architecture create a theater experience that exhibits the owners' preferences and is unmatched anywhere!

Installer: Media Systems, Boston, Massachusetts

Tuscan Sun

This theater was designed to mimic an old-world Tuscan-style veranda. The raised stage itself will ultimately be black carpeted. The stage kick will be faced with 2-inch thick x 8-inch tall x 6-inch wide stone that will match the faux finish of the columns/arches. This stone will also serve as a mason baseboard around the entire perimeter of the room – but only at the bottom of the darker back walls, not at the bottom of the columns and stem walls.

The Stewart microperfed Grayhawk screen is a very large 107-inch wide, 1.78:1 (16:9) with their patented ve-lux black velvet coated frame. Three of Triad's awesome Gold In Wall LCR/10's adorn the front screen wall, one in each stem wall that flanks the screen and stage and the center channel slightly "off center" behind.

The porch lights and stone baseboard lend the feeling of being in a courtyard. The rope lights help accent the columns and arches. The carpet does not call attention to itself and blends in perfectly with the other colors in the room.

All the flat ceiling tiles are absorption tiles and there are six pyramidal diffusion tiles in the ceiling. These along with the absorption wall panels on the front sidewalls (very difficult to detect – which is good!) create a room that is acoustically inert. This is not a "dead" room; it has all the proper elements of absorption and reflection. No echoes whatsoever! It is a fairly large room with a sizeable screen (107-inch wide, 123-inch diagonal). The projector is in a hush box in the upper right hand corner. The DVDs are stored discreetly in a decorative cabinet built into the back wall. It's movie time!

Installer: A-Plus Electronics, Wichita, Kansas

Viva Italia!

Named for a theater from the homeowners' dating years as teenagers, the Mayfair is a romantic Italian Renaissance style theater. Tasteful and elegant, it gleams with the richness of Old World charm and comfort, the details in 24 karat gold, and the Scalamandre fabric the same as that used for the Pope's throne in his last visit to the U.S.

Design by: Theo Kalomirakis Theaters, New York
Installer: Audio Command Systems, Westbury, New York

Ceiling Feeling

Bare bulbs add box-office feel to this spectacular home venue. Beautiful to look at, the tapestry-covered fiberglass panels that line the walls of this elegant theater also aid in low-frequency sound absorption. A Stewart screen receives its picture from a Runco DLP projector, while the room's audio gets a boost from a Meridian Dolby Digital/THX/DTX processor.

Installer: Audio Advisors, West Palm Beach, Florida

Super Sized

This 25-plus seat home entertainment complex is perfect for hosting dinner-and-a-movie theme parties. The two-piece projection system plays DVD, satellite, laser disc, cable TV, and VCR tapes on a 110-inch diagonal film screen. The audio system is digital and enhanced with Dolby and seven-channel surround sound. Everything in the room, including audio, video, curtains, temperature, and lighting, is controlled from a touchpanel remote control.

Installer: Audio Advisors, West Palm Beach, Florida

Togetherness

Although this is a small room for a theater (only 10-1/2 x 17-1/2 feet), the homeowner spared no expense, investing in a 9-inch CRT projector, Revel speakers, and mark Levinson amplifiers. This cozy setting with relaxed seating provides blockbuster movie viewing on the owner's schedule. The project cost $200,000.

Installer: Genesis Audio and Video, Irvine, California

Photography by Anthony Gomez

Photography by Anthony Gomez

Photography by Anthony Gomez

The makings for this home theater began with the homeowner's dream of renovating an unused space in the basement and transforming it into the ultimate getaway for enjoying movies and sports in comfort with family and friends.

The owner's specific criteria were as follows: to create a state-of-the-art home theater environment, a high-end acoustic experience, and an innovative lighting system, all of which would be managed by easy to use touchscreen and keypad programming.

Achieving the owner's goals was not without certain design challenges. In particular, the ceiling height, acoustic isolation, and ventilation of the existing space required some innovative design solutions.

The design process started with the decision to seat up to eight adults comfortably for viewing movies and sports on a large wide-screen display. In addition, four smaller 24-inch displays would be available beside the main screen for monitoring multiple sporting events or any of the CCTV cameras located on the property. These four displays conveniently disappear behind motorized black fabric panels when not in use to avoid detracting from the beautiful image on the main screen.

Video quality was very important, so a SIM-2 SVD-800 Millenium CRT projector was partnered with a Faroudja processor and the image is displayed on a Stewart ElectriMask screen. This combination of SIM-2's finest projector (calibrated to ISF standards) and an excellent processor and screen is truly breathtaking. The black ceiling and front wall cause the main image to really jump off the screen with great contrast and vibrant colors. A total of four satellite receivers and two VCRs allow virtually any program to be enjoyed simultaneously on the multiple screens. An extensive DVD library is partially stored in Escient's PowerPlay DVD manager for easy and convenient retrieval of any movie that is loaded.

To keep pace with the image, a first class audio system was built around a Lexcion processor and power amplifiers, with speakers from Paradigm's Reference line including two 15-inch servo-controlled subwoofers. Three identical front speakers and the two subs are concealed behind grille cloth under the main screen. Dipoles are mounted to the sidewalls and matched in-ceiling

speakers are mounted above the back row in custom sealed enclosures. All speakers use identical drive units for timbre matching.

In spite of the tight confines, three full-height Middle Atlantic AX-S slide-out racks were installed at the back of the room. The racks are hidden behind a set of double doors that have been treated with the same acoustic panels as the rest of the room for a very discreet and quiet installation. A 15-inch LCD Monitor mounted in the rack permits easy monitoring of any source and the ability to preview the viewer's selection before sending it to one of the five displays on the front wall.

Pulling everything together is an Elan VIA! touchpanel. Located on a console beside the owner's primary seat, the touchscreen makes it easy to activate and control every aspect of the home theater, including video source selection and control, audio format, room temperature, and the Vantage lighting system.

Installer: Audio Video Unlimited, Coquiltam, British Columbia

Center Stage

Everyone deserves the chance to rock and roll now and then, and what better place than your own private home performance stage? These homeowners wanted to capitalize on the entertainment space by making it versatile. With connections for amplifiers and other auditory enhancers, this space goes from home theater to amphitheater with ease! This is also an optimal practice area for a young hopeful band, who wishes to graduate from the cacophonous acoustics of a garage. Or, for those who love to perform but want a little backup, the space is perfect for a karaoke machine to enhance the vocal experience, especially with the large screen behind for lyrics or other visual accompaniment.

Installer: Home Media, Ketchum, Idaho

Original Artwork

It is becoming more common that the woman of the house plays an active role in the decision making process of purchasing a home theater. On this project, which received the award for Best Home Theater, Level II – $100,000-$125,000 in 2001, the wife was completely in charge. As an artist, her eye was better than most videophiles. She looked at many different technologies before settling on a 9-inch CRT projector as producing the most film-like image. To match the video quality, a Lexicon processor was used with nine Snell speakers and two of NHT's flagship subwoofers. The system controller makes the system easy enough that even the youngest child in the family can operate it.

The homeowner expressed her preference for a more modern look than the traditional cinemas she had seen in magazines. Space for her artwork (all her own work) inside the theater and lobby was also a requirement. It was a bit of a challenge to design a modern look while at the same time placing acoustical treatments without creating too "soft" a look. By constructing absorbers and diffusers that doubled as architectural elements, the typical "fabric wall" look was avoided. Finishing touches were applied with lighting: uplights on the columns, bullnose step lights, and a lighting scene controller.

Installer: Silver Screen and Sound, Towson, Maryland

Equipment List

1 Elite Video DH-4600 9-inch CRT front projector
1 Progressive ceiling mount for projector
1 Draper Cineperm 133-inch diagonal 16:9 fixed mnt (10' wide)
1 DWIN TranScanner 2 Variable multiplier
1 Lexicon MC-1 7.1 Channel THX/DTS/Dolby Digital EX processor
1 Anthem MCA-2 II Front channel amp, 2 x 200 watt
1 Anthem MCA-5 II Center/Surround amp, 5 x 200 watt
1 Marantz DV-4000 DVD/CD player
1 Philips VR999 S-VHS VCR, Hi-Fi
1 Marantz ST-6000 FM/AM Tuner
1 RCA DTC-100 HDTV DSS receiver and elliptical dish
3 Snell XA 1900 Left/Center/Right speakers
2 Snell SR.5 Side omni-directional speakers for 1st row
2 Snell SR.5 Side omni-directional speakers for 2nd row
2 Snell SR.5 Rear omni-directional speakers
1 NHT Sub 2 Powered subwoofer w/dual 10-inch drivers and 500w Sunfire amp
1 NHT sub 2S Slave subwoofer with 500 watt Sunfire amp
1 niles IPC-8 System surge protector
1 Monster – Audio/Video interconnect package
1 Niles Intellicontrol Ultra-simplified system controller
1 Lutron Grafik Eye Room lighting scene controller
10 Leather Center Magellan Leather cinema recliner w/stainless steel cupholders
6 Bass Presidential Cinema seat, rocking, high back (rear row)
1 SSS – Various acoustical treatments, both absorbing and diffusing

Photography by John Coyle Jr., Baltimore, Maryland

Photography by John Coyle Jr., Baltimore, Maryland

Architectural Allowance

When it came time to design their new custom built home, these Toronto-area homeowners knew there was to be a space allotted for a dedicated home theater. The architect understood their needs and the result was a 16 x 30-foot space in the lower level of the home, complete with height allowances for a finished ceiling height of nine feet and a custom designed HVAC system. Further, the floor of the foundation was modified to allow for tiered seating. A double door and adjacent, secured space for audio/video equipment facing into the theater were also provided. A tremendous asset to the project was an established custom homebuilder who paid close attention to the entire project and was always available to handle any concerns or bumps in the road.

Installer: Howell & Associates, Inc., Burlington, Ontario, Canada

CRESCENT BIJOU

Nose to the...

Greeted by an elaborate Art Deco design as they approach this home theater, guests prepare to step into the past as they enter this replica of a 1930s film studio screening room. A Runco DTV-1101 data-grade CRT projector provides superior picture quality, while M7K THX speakers support the images with superior sound. The Art Deco-style chairs from Acoustic Innovations are outfitted with a Crestron VT-3500 touchscreen for complete control.

Installer: Howell & Associates, Inc., Burlington, Ontario, Canada

MERIDIAN

Deco Echo

Decorated in a flashy Art Deco style, this theater was designed to re-capture the dazzling glamour of early to mid-twentieth century Hollywood.

Installer: AVL Pro, Inc., Naples, Florida

Reference Room

The goal for this theater project was to build a home theater in a pre-existing structure, using the highest level of components for the owner's requirements. The projector selection was a Runco VX-3 DLP projector, housed in a room directly behind the theater in order to create an open, spacious feeling. It also eliminated fan noise.

The final equipment choices included Vandersteen 5 Reference Loudspeakers for two-channel listening, as well as for front theater speakers. Three pairs of VSM signature surrounds and Vandersteen's VCC5 center channel complete the speaker configuration. Four Aerial SW12 subwoofers provide the maximum amount of low-end information.

A 100-inch Stewart Filmscreen was chosen to realize full projector potential. Mark Levinson's ML40 Reference Surround Processor and seven sleek ML436 monoblock amps for center and surround speakers were selected as central components. The primary A/V source component is the Meridian 598 DVD-Audio player. The theater is operated by a Crestron ST-1550C remote control. Interconnects are Transparent Reference with Transparent Opus speaker cables. All equipment is displayed in four 8-foot tall Middle Atlantic equipment racks outside the theater.

Installer: Audio Advice, Raleigh, North Carolina

Back Bar

The look of this theater is one of classic, traditional elegance, using dark grays and greens in the color scheme and also employing richly textured wood trim. The front speakers are small monitors from Aerial Acoustics. Along with the Aerial center channel, they are concealed behind grill cloth, directly beneath the 100-inch Stewart Filmscreen. Aerial surround speakers are on side walls, with smaller B&W in-wall speakers mounted in the limited area on the rear wall behind the bar. A small B&W subwoofer was also placed in the rear of the theater.

A Runco CRT projector provides the images for the screen, and a Sony 20-inch television is mounted to the right of the screen to enable monitoring of the front door or to play alternate programming or sporting events. The theater's front B&W subwoofer is concealed beneath the smaller TV.

System components with the need for immediate access are located in a cabinet behind the bar. These include a Lexicon MC-1 processor, a Faroudja line doubler, and a Sony ES DVD player and VCR. The home's remaining home automation products are hidden beneath the wall in a separate equipment room with access on the wall's other side. The theater is controlled by a Phast Viewpoint color touchscreen, which also controls the home audio system and a Lutron lighting control system.

Installer: Audio Advice, Raleigh, North Carolina

The mission was to create a technologically advanced, easy-to-use home theater, with bright, clear images and first-class sound in a room originally intended to be a recreation room and bar. This 17 x 17 x 8-1/2-foot space provided a number of challenges once the owners decided to make it a dedicated home theater. The first was to overcome the acoustic problems of the room shape. All of the wall and ceiling surfaces were constructed using ASC's Walldamp Iso Wall system to help even bass response and minimize sound transfer between the theater and surrounding rooms.

The cabinetry was designed in coordination with the architect to provide space for the home theater equipment, a complete twenty-five zone house music system, supplemental equipment ventilation, proper screen position, and adequate space behind the acoustically transparent screen for the front speakers and subwoofer. Generous media storage is also provided by full height roll-out shelving adjacent to the equipment racks.

The 1.33:1 screen ratio was chosen because the owners anticipated watching sporting events and classic movies, as well as newer widescreen DVDs. Drapery along the entire back wall blocks light from the windows, and softens the acoustics at the back of the room. The over-sized custom sofa offers a commanding view of the show, as well as impeccable comfort.

Control of the system is fun and efficient with the AMX Viewpoint remote control system. A few intuitive presses and even houseguests are taking command as the lights dim and the show begins. In addition to simplified control of the home theater and Lutron dimming system, the touchscreen can control the house music system or show security camera images on the main screen.

Installer: The Whole Ball of Wire, Inc., Glenwood Springs, Colorado

CINERAMA
South Seas Adventure
GRAMMY 2000
VALET PARK

Critics' Panel

Viewers can make their own judgment about the film at hand in this contemporary home theater, where formal wood paneling vaguely resembles a courtroom! Swanky studded leather chairs and wall-to-wall Oriental-style carpet make it a little more like home. Speakers mounted on either wall provide the ultimate surround sound. A home office area is housed with the theater electronics.

Installer: Advanced Residential Systems, Allentown, Pennsylvania

Photography by John P. Hamel

Photography by John P. Hamel

Photography by John P. Hamel

Keeping the Kids Close

Now that his children are teenagers, this homeowner was looking for a way to remodel their playroom into something that would keep them at home. The result was a small yet intimate home theater with high performance quality. The powerful but easy-to-use system retains a look of elegance.

B&W Nautilus 801 speakers, powered by Mark Levinson amplifiers are fed by sources such as CD and DSS audio utilizing digital audio outputs. The four-way Stewart screen is lit by a Runco high-definition projector fed by DSS and progressive scan DVD.

Acoustical treatment consisted of absorptive panels in the front, while the rear wall utilized RPG Diffuser panels. Twelve zones of lighting were integrated into the AMX Landmark system and controlled by wall-mount keypad as well as wireless touchscreen.

In order to accommodate the family and their friends, six Irwin Ambassador recliners in black leather were incorporated for a comfortable, yet stylish solution to theater seating. Behind the two rows of seating is a sit-down bar with comfortable stools, complete with VGA and Ethernet connections for broadband surfing via laptop computer.

Installer: Media Design, Inc., Houston, Texas

panja.

Crawl Space Expansion

Meticulous attention to acoustic principles, combined with Genelec surround sound speakers and a Sony CRT projector, turned an unfinished basement into a studio-quality home theater. During the basement's construction, a 23 x 14-foot theater was created. Designed with cinderblock construction and a 3-foot wide hallway separating it from other areas, the theater enjoys sound isolation from the rest of the basement. Originally, the basement was a 5-foot high crawl space with a sand floor under an existing structure. Getting digging machinery there was not an option, so the construction crew had to dig twelve feet down by hand.

Acoustic and architectural design was a collaborative effort with a prominent design group, and studio features were incorporated into the residential setting. A separate parametric equalizer controls each speaker so that frequencies for each channel can be adjusted. The 9-inch Sony CRT Projector/Faroudja scaler fires on to a Stewart ElectroMask screen that adjusts to various aspect ratios, producing a true cinema quality picture. All components and setting adjustments are controlled from the custom touchscreen layout.

Installer: Media Systems, Boston, Massachusetts

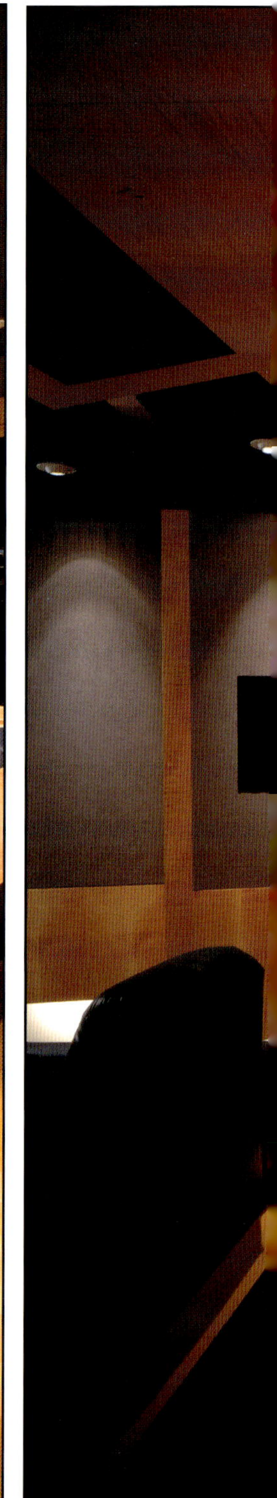

Lounge Wizards

The owner of this home theater wanted to design a room that housed the highest quality audio and picture available. The catch: the screen, projector, and speakers must be out of sight when the system is not being used. The solution: to put a Runco DTV 1101 projector and a Stewart Mirror System in its own projection room, located behind the screen. The screen itself is hidden, when not in use, by automatic, motorized doors. A CAT in-wall speaker system, fifteen speakers in all, was built into the walls and hidden behind custom-made acoustic paneling.

The seating was designed to be casual and comfortable for the family of seven. Formal "theater style" seating was set aside for plush sectional sofa seating that can accommodate as many as twenty-five guests.

Controlling the system is a cinch with two Crestron touchpanels located in the room. One touchpanel, a 12-inch wall-mount model, is located next to the rack, while a wireless STX 1550c lets the viewer lounge anywhere in the room and still have control. Want to watch a movie? You don't even have to put the DVD in the player. The Escient Powerplay DVD jukebox can be accessed from each touchpanel. Any DVD in the collection can be selected by touching the cover art of the movie, which is displayed on the touchpanel. Once selected, the lights dim, the screen is revealed, and the show begins!

Installer: Audio Visions, Woburn, Massachusetts

The goal for this project was to create an alluring theatrical environment without resorting to the traditional ornamentation used in the old movie palaces. The Roxy had to blend in with the style of the main house, which combines contemporary architecture with streamlined modernistic touches. The goal was accomplished successfully with creative use of architecture, lighting, and interior design.

One of the most unusual characteristics of this theater is the abandonment of the classical symmetry that defines the old movie palace. Each one of the four walls here has its own character. The left wall features an exciting sculptural arrangement of millwork elements that incorporates concealed absorptive acoustical treatments. The right wall is defined as a series of columns that separate the theater from the lobby next to it. The rear wall creates a sculptural effect mixing lighting, wood, and acoustical paneling. The upper part of the theater is framed with a frieze that features Frank Lloyd Wright-inspired, rear illuminated panels. Finally, the ceiling completes the picture with a multi-layered arrangement of acoustically treated, cove-lit surfaces.

Special attention was paid to the room acoustics. To make sure that the performance of the sophisticated audio/video system was not compromised, the acoustical consultants of the project carefully defined the exact relationship between the height, the width, and the length of the room and made sure that the walls were treated with the appropriate reflective, absorptive, or diffusive acoustical panels – all concealed behind the architectural design.

By paying equal attention to technology and architecture, the team responsible for the Roxy succeeded in creating a space that combines the exuberance of movie palaces from the thirties with the sophistication of contemporary design and technology.

Installer: Integrated Media Systems, Sterling, Virginia/ Theo Kalomirakis Theaters, New York, New York

This theater won CEDIA's Electronic Lifestyles contest for Best Home Theater, Level I – $45,000-100,000 in 2001.

The homeowner asked for a "southwest themed getaway," and wanted the projected completed in less than a month. Additionally, the room was surrounded on all six surfaces by concrete, causing a tremendous echo problem. The timing problem was overcome by a concerted effort between the trades involved and by efficient project management. The echo was tamed by upholstering the walls between the columns and behind the screen area. Since the owner wanted an old Southwest feel to the room, it was suggested that shutters from an old farmhouse be installed to serve a decorative function and to double as sound diffusion panels. Reclining cinema chairs complement the décor of the room. The project was completed on time, sounds great, and has become the owner's favorite part of the house.

Installer: Audio Video One, Winston Salem, North Carolina

Equipment List

Runco DTV-863 multiple aspect CRT projector
Runco DTV-863 video processor
Stewart Filmscreen 110-inch diagonal Luxus Deluxe Screenwall
Atlantic Technology 370 THX Ultra speaker system
Crestron ST-1550c wireless color touchpanel
Crestron ST-CP control processor
Crestron ST-10 input/output module
Crestron ST-CS current sense module
Crestron ST-IRP infrared probes (4)
Integra DTR-7.1 A/V receiver
Integra DPC-6 6-disc DVD carousel
RCA DTC-100 satellite receiver
Panasonic PV-VS4820 S-VHS
Extron SW2ARHVxi RGB switcher
Lutron GRX2403A Grafik Eye lighting control
Panamax Coaxmax6 surge protector
Panamax Coaxmax2 surge protector
Panamax MaxDBS+ surge protector
Proflex Interconnects
Extron RGB and RS-232 cables
Acoustic Innovations Cinemachairs
Bass movie posters

"THE SHEIK"
GROUCHO
CHICO
HARPO
MARX
BROTHERS
A NIGHT AT THE OPERA
KITTY CARLISLE

Visual Connection

This magnificent home theater was a dream-come-true project in many ways. The very first meeting with the homeowners, architect, and builder resulted in a clear understanding that the theater was actually the major impetus for the renovation of the entire basement. The purpose of the renovated space would be to entertain friends and family, particularly during the NFL season. "Let there be no room in this space where we won't be able to see and hear the game" became the marching orders. As a result, the audio and video playing in the theater was made available to a 42-inch wall mounted plasma TV at the bar, and a 12-inch LCD panel in the bathroom! As for the theater, it was clear from the beginning that no expense would be spared to make the room and the screen as large as possible. Given the mandate to fill a 13-1/2-foot wide screen with superb video and to provide high fidelity, high SPL audio for the 6,500 cubic foot space, it was immediately decided to utilize the Runco VX-3c 3 chip Runco DLP projector and a JBL Synthesis One package.

Initially the challenges were physical: the theater's desired length required the architect and builder to "blow out" a foundation wall and excavate under an existing garage. Acoustical challenges were solved with discreet placement of absorptive and refractive RPG diffuser panels, all covered with beautiful fabric chosen by the interior designer. After overcoming many installation delays and minor setbacks, the project was finally ready for the homeowner to entertain his family on Christmas Eve.

Installer: Integrated Media Systems, Sterling, Virginia

Homey Atmosphere

This room is a dedicated theater with a warm, comfortable feel. The SVS mini motorized scissor lift allows the projector to be hidden away when it is not in use. Source gear for the room (and the rest of the house) is hidden behind curtains located on either side of the screen.

Installer: Designer Audio, Naples, Florida
Builder: Gulfshore Homes

Photography by Dean Callis

Photography by Dean Callis

Eye Scream

Ice cream parlor décor was used in the lobby area of this space-luscious theater area. Inside, tiered, super-comfy leather loveseats allows for snuggling or single-seat reclining.

Installer: Silver Screen and Sound, Towson, Maryland

Photography by John Coyle Jr., Baltimore, Maryland

Photography by John Coyle Jr., Baltimore, Maryland

This theater room, from its conception, was designed with performance as the primary objective. Early involvement with the architect allowed for careful planning of room dimensions to allow dramatic "stadium-style" seating and complete acoustical modeling and placement of room treatments.

The theater system is based on the large Runco 9-inch CRT projector utilizing a 9 x 5-foot multi-aspect ratio Stewart filmscreen. Audio is provided via Revel Ultima series speakers and subwoofers used in conjunction with Sonance THX in-wall surround speakers. The audio system is comprised of Proceed electronics, including seven channels of brute-force 250 watt/channel amplification.

The room's walls are covered with black acoustically transparent fabric to cover the treatments behind them. Speakers in the front of the room are hidden behind the same fabric, while the large powered subwoofers are concealed with large Reggio designer cast iron grilles, usually used to cover heating or air conditioning ducts. The architect and the owners insisted on the room having four operable windows. This was a challenge, as the windows presented problems with ambient light and acoustical properties. Modern Home Systems devised motorized panels that slide in tracks to cover the windows automatically when the theater is activated. The panels then retract into pockets when the system is shut down, giving the room the natural light and air circulation the homeowners desired.

Modern Home Systems designed and installed a complete Crestron home automation system within the home to control the theater, four additional surround sound "mini-theaters," and other low-voltage systems. The net effect is an extremely high-tech modern home that is controlled and managed in a very simple way.

Installer: Modern Home Systems, San Diego, California

Fantastic Five-seater

Exotic carpeting and lush leather chairs provide a luxurious viewing experience. The rich wood paneling aids in the acoustics of the room, and black panels conceal speakers and other components.

Installer: Audio Visions, Inc., Lake Forest, California

Full Tilt

This contemporary home theater provides casual comfort for the movie enthusiast. Two tiers of seating are available for family and friends to convene, with three leather recliners making up the front row, and an exotic jungle-printed couch that gives viewers an opportunity to sit or lie down while enjoying the current feature. The unique speaker configuration provides optimum sound, and the Art Deco-style black and white décor, including the geometric fixtures on the wall, suggests a classic, yet modern atmosphere.

Installer: Silver Screen and Sound, Baltimore, Maryland

Photography by John Coyle Jr., Baltimore, Maryland

Photography by John Coyle Jr., Baltimore, Maryland

Room to Grow

Although their builder had suggested a theater for this space in their 5,000 square foot home during the preliminary construction planning, the homeowners chose to wait a year before going ahead with the project. When they were ready, they went back to the builder to ask for a referral. He had felt from the beginning that a theater would be just perfect for them, because they love movies and entertaining.

The theater space dimensions were 16 x 24 x 9.5 feet. So that all viewers could comfortably enjoy the show, soft leather reclining chairs were selected, arrayed in three levels of four seats each. The room was treated with staggered stud construction and Acoustic Science Corp.'s Deci-Damp sound absorption system. The interior walls were treated with acoustical panels. A 110-inch Stewart Wide Screen was the perfect size image for the cozy room, and a Sony G90 CRT video projector and JBL Synthesis II Surround EX system delivers a superb video and audio experience. To make system operation easier, a Panja Viewpoint control system was installed, allowing theater occupants to control the theater by simply touching menu items and icons. In addition, a six-zone Lutron GRAFIK Eye system was installed to handle the lighting control. For the owner's constantly growing collection of movies, a Sony 200 Disc DVD jukebox was incorporated along with plenty of hidden storage space for the DVD movie library. Last, for a complete hospitality experience, the builder designed a wet bar in the rear of the room for refreshment service.

Installer: Electronics Design Group, Inc., Piscataway, New Jersey

© 2003 Electronics Design Group, Inc.

Sound Lounge

This custom home theater features acoustical details and mohair covered "Royal" loungers by Acoustic Innovations. The custom enclosed and self-cooled projector enclosure, which reduces noise to a minimum, houses a Sony G 90 CRT projector. The picture on the Stewart 110-inch Luxus screen is enhanced with a Faroudja DVP 5000 and is specially calibrated to deliver precise, stunning images. Audio processing is delivered through a Kenwood Sovereign Receiver and Parasound Amplification delivers breathtaking sound through the Atlantic Technology 450 THX speaker system. Clark Synthesis Transducers enhance the thunderous low frequency effects (LFE) from action movies.

Installer: Monaco Enterprises, Inc., Spokane, Washington

Glossary of Techie Terms

CRT projector. A projection system that uses three separate picture tubes each connected to a series of lenses. There is a specialized CRT/lens assembly for each color: red, green, and blue.

db (Decibel). A logarithmic unit of measure commonly used to gauge relative volume in terms of dB SPL (decibels of Sound Pressure Level). The decibel is used as a standard unit for expressing the ratio between power and power level.

DBS (Digital Broadcast Satellite). High-powered, high-band digital satellite TV transmission system using a dish that measures just 18 inches in diameter.

Defractor. In terms of photographic lenses, used to add a ray-casting prism/rainbow effect to light sources in a picture or movie.

DLP (Digital Light Processing). A type of high-resolution front and rear projection TV system invented by Texas Instruments in the mid 1990s. A DLP projection system consists of a light source, up to three tiny digital micro-mirror devices, a projection lens, and a cooling system. It is capable of projecting bright, movie theater-sized images.

DSS. Digital Satellite System.

DVD. Digital Video Display.

Electrostatic. Sound created using tall, thin speakers that use a membrane to emanate a unique, articulate, and very accurate sound.

HDTV. High Definition Television.

HVAC. Heating, Ventilating, and Cooling.

Interlaced NTSC to 480P. In terms of signal conversion, also known as de-interlacing. Every video scaler must first de-interlace the signal before it can process it. This is the most important step in the scaling process; if it is done poorly you will never get a good quality picture. It is the process where the signal is converted from alternating "fields" of 240 lines every 1/60th of a second, which lay on top of each other to make one full "frame" every 1/30th of a second, to full "frames" of 480 lines every 1/60th of a second. It was originally termed "line doubling" in the early days when Faroudja was the only game in town. In terms of input compatibility, Interlaced NTSC (a.k.a. 480i) is the most basic level in terms of picture quality. All Cable TV, VHS, Laserdisc, DVD, and standard definition satellite signals are "Interlaced NTSC." 480P (a.k.a. 480 Progressive) is the next step, typically delivered via Component or RGB video and is commonly output from Progressive-scan DVD players and some HDTV receivers.

LCD (Liquid Crystal Display). A type of display that utilizes two sheets of polarizing material with a liquid crystal solution.

LCR. An abbreviation/acronym for the front three speakers (left, right, center) in a home theater sound system.

Line tripler. This technology monitors the incoming picture and then reproduces the picture with three times as many lines of information. These extra lines decrease the jagged edges and improve the richness, or texture, of the image, especially with large screen sizes.

LFE. Low frequency effects.

Plasma. A flat-screen monitor that uses thousands of sealed, low-pressure glass cells filled with a mixture of neon and xenon. Behind each cell are colored phosphors, one red, one blue, and one green. When energeized, these cells of plasma emit invisible UV light that strikes the red, blue, and green phosphors on the black glass display, making them produce invisible light. These monitors are approximately 3-5 inches thick.

Processor. In terms of video; a scaler, line doubler, quadrupler, or color decoder.

Proscenium. In a modern theater or similar buildling, the part of the stage in front of the curtain, sometimes including the curtain, its arch, and the nearby ceiling.

PVR. Personal Video Recorder, such as Replay or TiVo, which uses personal computer hard drive and internal processing to drastically increase recording time, compared to standard VCRs, allows "smart" programming, and also allows pause control over live broadcasts.

RF two-way. Refers to the bidirectional communication between the handheld remote control (two-way RF touchscreen) and the system control processor. This means that not only does the touchscreen tell the system what to do, but the processor can send feedback to the two-way RF remote control to let the user what is going on in the system.

THX sound. A set of patents, technologies, and playback standards for reproducing film soundtracks at home.

Touchpanel/Touchscreen. A visual display terminal screen that responds to instructions and icons as the user touches the screen.

Transducer. A device that converts one type of energy (electrical, acoustical, magnetic, or mechanical) into another form of energy. Speaker drivers, microphones, and phone cartridges are examples.

VHS. Dominant home VCR (video cassette recorder) standard.